MEDICAL CONSEQUENCES of the CHERNOBYL NUCLEAR ACCIDENT

MEDICAL CONSEQUENCES of the CHERNOBYL NUCLEAR ACCIDENT

Edited by

P.V. RAMZAEV

NOVA SCIENCE PUBLISHER, INC.

Art Director: Maria Ester Hawrys
Assistant: Director: Elenor Kallberg
Graphics: Denise Dieterich and Kerri Pfister
Manuscript Coordinator: Roseann Pena
Book Production: Tammy Sauter, and Benjamin Fung
Circulation: Irene Kwartiroff and Annette Hellinger

Library of Congress Cataloging-in-Publication Data

Medical consequences of the Chernobyl/ edited by P. V. Ramzaev.
p. cm.
Includes bibliographical references and index.
ISBN 1-56072-111-1 :
1. Chernobyl Nuclear Accident Chernobyl', Ukraine, 1986--Health aspects.
I. Ramzaev, P. V. (Pavel Vasil'evich)
RA569. M43 1993   93-22794
616.9' 897'00947714--dc20   CIP

© *1996 Nova Science Publishers, Inc.*
6080 Jericho Turnpike, Suite 207
Commack, New York 11725
Tele. 516-499-3103 Fax 516-499-3146
E Mail Novasci1@aol.com

All rights reserved. No part of this book may be reproduced, stored in a retrieval system or transmitted in any form or by any means: electronic, electrostatic, magnetic, tape, mechanical, photocopying, recording or otherwise without permission from the publishers.

Printed in the United States of America

CONTENTS

Introduction 1
P.V. Ramzaev

The Level and Structure of Morbidity in Children Living on the Territories of the Ukraine Affected by Radioactive Contamination as a Result of the Chernobyl Accident by 5
V.N. Bugaev, T.V. Treskunova, and E.I. Bomko.

Medical Consequences of the Radiation Accident in the Southern Urals in 1957 15
L.A. Buldakov, S.N. Demin, V.A. Kostyuchenko, N.A. Koshurnikova, L.Yu.Krestinina, M.M.Saurov, I.A. Ternovskij, Z.B. Tokarskaya and V.L. Shvedov

Consequences of Chronic Radiation Exposure 29
L.A. Buldakov and A.K. Guskova29

The Experience of the First Nuclear Facility (Exposure Levels and Staff Health) 39
B.V. Nikipelov, A.F. Lyzlov, and N.A. Koshurnikova

Changes in Reproductive Functions of Mice Induced by Isolated and Combined Action of X-Ray Radiation and Stress, 53
V.L. Vaskan,

Oncologic Morbidity and Mortality in Areas Under the Control of the Brjansk Province Following the Chernobyl Nuclear Power Plant Accident 61
A.A. Dudarev, G.I. Miretsky, P.V. Ramzaev, M.N. Troitskaya, and I.E. Shuvalov

Cytogenetic Effect in Somatic Cells of Persons
Affected by Radiation Exposure in Connection
with the Chernobyl Accident   75
    M.A. Pilinskaya, A.M. Shemetun, A.J. Bondak, and
    S.S. Dybskij

Hygienic Evaluation of Thyroidal Radiation Doses
in Inhabitants of the UkraineFollowing the
Chernobyl Atomic Power Station Accident   83
    A.Ye. Romanenko, I.A. Likhtarev, N.K. Shandala et al.

Health Status of the Adult Population in the
Western Districts of the Bryansk Area in 1989   93
    R.N. Turaev

The Impact of Low-Dose Ionizing Radiation
on the Progress and Outcome of Pregnancy
in Women   109
    O.S. Ul'yanova, N.I. Mashneva

Evaluation of Leukemia-Induction Risk Based
on Analysis of the Consequences of Nuclear
Incidents in the Southern Urals   133
    M.M. Kosenko, M.O. Degteva, N.A. Petrushova

Subject Index,   133

INTRODUCTION

It is the author's belief that thesearticles analyzing the medical implications of the Chernobyl accident will be important among the increasing amount of works that are devoted to the problem that have been published in various countries.

The authors of the papers represent various scientific institutions of the former USSR: they were among the originators of truly scientific, strictly impartial information concerning the medical implications of the Chernobyl NPP accident. It is no secret that the flurry of publications appearing on the heels of the accident, penned by both foreign authors and those of "the one-sixth of the world", were free of either hard facts or competent analysis. Among those out for blood, noted writers and journalists were the most ferocious. They were hell-bent on taking over the highest ranks of authority by using the pretext of their "concern for the people". They stubbornly refuted genuine scientists by their purely fictional exercises. One may recall the hysterical public speeches by some Ukrainian and Belorussian writers and the attempts of certain authorities to sue medical scientists for their alleged deliberate underestimation of the accident's consequences.

Under pressure from these "truth champions", the USSR government virtually expressed a lack of trust in Soviet medical science by appealing to IAEA for help in an objective examination of the situation in the contaminated zone of the three republics. The results of the examination are now universally known, due to the reports presented at the IAEA conference (May of 1991). It appeared that instead of covering up the consequences, (official classification of the data as secret has been enforced by the authorities rather than scientists), the Soviet scientists overestimated them at the very least, judging by the irradiation doses, (by a factor of 2). Factual data concerning the health condition, reported for the first time in the papers of this Collection, fully agrees with the conclusions reached by the IAEA conference held in May of 1991.

People will always remember the irretrievable losses of the accident, 145 people falling prey to radiation sickness. 610 people have received doses exceeding the permissible emergency "standard" (25 ev). The trial has named the guilty parties and determined the extent of their blame.

Presently, the subsequent turn of events and society's response to the Chernobyl accident are causing is having an impact which is hundreds times worse than that of the direct losses suffered as a result of the accident. The collective dose expected to be received for 70 years by the population of the three republics, Russia, Ukraine and Belarus, amounts to 0.2 to 0.3 million people ev. According to universally adopted estimation techniques, the justified expenditures of eliminating the dose will cost 2 to 3 billion rubles (in 1989 prices). On the other hand, the cost of protective measures that reduced the dose approximately by half has already exceeded the quoted figure many times over. An estimate exists suggests that already implemented measures have cost more than 100 billion rubles. Operating programs and laws call for even more increase in expenditures without regard for their efficiency. And these events take place in poverty stricken CIS countries that are suffering from a lack of resources capable of supporting the life and health of the millions of people suffering from obvious causes.

This collection's contributors firmly uphold the ICRS position which maintains a threshold-less concept of stochastic effects and requires the reduction of the dose to levels feasible under the existing social and economical conditions. This existing risk of stochastic effects induced by the radiation dose received by the population is so far too low to be determined from the cases of spontaneous carcinogenesis in the contaminated areas during the time passed since the accident; only a rise in the chromosome aberrations rate has been detected so far, and the health implications of the latter are uncertain. While assessing the deterministic effects, one should note that the experience gained by the scientific world in examining millions of people is a strong argument against this possibility unless the yearly chronic dose exceeds 10 ev, (critical level for lenticular capacity), and the acute whole-body doses exceed 50 ev (radiation sickness). Similarly, damage to the thyroid gland's functions is detected when the doses exceed 10 Gy.

It is a matter of a common knowledge now, that those damaging levels of radiation were detected outside the nuclear power plant territory, in some limited spots of Pripyat town, (about 1 Gy), and in the "red forest" area. Doses exceeding 10 Gy on the thyroid gland have been detected for a number of children, (2000 people in all, probably), that had been drinking milk and failed to take stable iod. Classifying the rest of the population outside the mentioned areas as affected by radiation is unjustified, and apparently, in itself, can lead to additional health problems for the population. This classification should be applied only to those who suffer or could have suffered radiation damage and need or have needed medical treatment. No people fitting such a description have been found outside the 30-kilometer zone.

The question is: "Why should both domestic and foreign science go to all these lengths in analyzing the health impact of Chernobyl?". The efforts can hardly be explained by a rank curiosity or a pursuit of the worldly goods on the part of the scientists. Indeed, there exist powerful reasons for such large-scale research. Ionizing radiation is such a global and ever present factor, especially in this nuclear age, that it could not be removed from the biosphere by magic. Thus, detailed knowledge of its low-dose effect on human health, metabolism and gene pool becomes a matter of extreme importance for radiobiology and hygiene. Despite a great supply of data on biological implications of radiation exposure, scientists are still not altogether sure that radiation levels caused by the accident are absolutely safe. And although the accident at the Chernobyl NPP was not the first, it is by far the most dangerous and, unfortunately, it is not going to be the last in this sad line.

Five years later, a fire broke out once again at the same power plant; a third-degree emergency occurred at the St. Petersburg NPP. Knowledge of medical implications of the accident, especially remote ones, will form the basis for making optimum decisions involving all three branches of radiation hygiene: analyzing the process of the dose-forming in humans, assessment of the doses health impact and the development of a safe and reliable system of radiation safety.

It was the intention of these scientists to make these reports a contribution, albeit a modest one, to this noble endeavor of mankind.

Member of Committee-I ICRS,
Director of the St. Petersburg
Institute for Radiation Hygiene,
Doctor of Medicine, Professor   P.V. Ramzayev

THE LEVEL AND STRUCTURE OF MORBIDITY IN CHILDREN LIVING ON THE TERRITORIES OF THE UKRAINE AFFECTED BY RADIOACTIVE CONTAMINATION AS A RESULT OF THE CHERNOBYL ACCIDENT

V.N. Bugaev, T.V. Treskunova, and E.I. Bomko
Ukrainian Scientific Center of Radiation Medicine, Kiev

A study of the health status of critical population groups, (pregnant women, newborns, children and teenagers), of the territories contaminated as a result of the Chernobyl accident [1-4,7] is of special importance. Surveillance of critical subpopulations for the purpose of development and specifications of a medico-biological concept of prolonged, (commensurate with man's lifetime), residence of the population in the contaminated area without prejudice to health and disturbance of a traditional way of life is one of the most topical tasks of contemporary life. Active detection of individuals with functional and pathologic changes in health status suggests an examination of the population, a long study of representative groups, a choice of individuals at risk of developing various disease or other unfavorable consequences of ionizing radiation.

As a result of the accident, radioactive contamination of a territory has taken place in the Ukraine. The main dose-forming radionuclide at the oral intake of a substance into the human body if Cs-137. The largest density of Cs-137 contamination was found in the Luginsky, Narodichi and Ovruch districts of the Zhitomir region and in the Ivankovsky and Polessky districts of the Kiev region, which are ascribed to the so-called

Table 1. Radioactive contamination and the size of the child population in the controlled areas of the Ukraine

District	Cs-137 contamination density*, Cu km^2	Number of settlements	Size of child population, thousands
Luginsky	4.5	38	1.025
Narodichi	11.5	68	4.553
Ovruch	3.56	81	16.2
Ivankovsky	1.7	40	6.8
Polessky	8.5	47	6.084

*According to the data of the State Committee on Hydrometeorology as of July 1989.

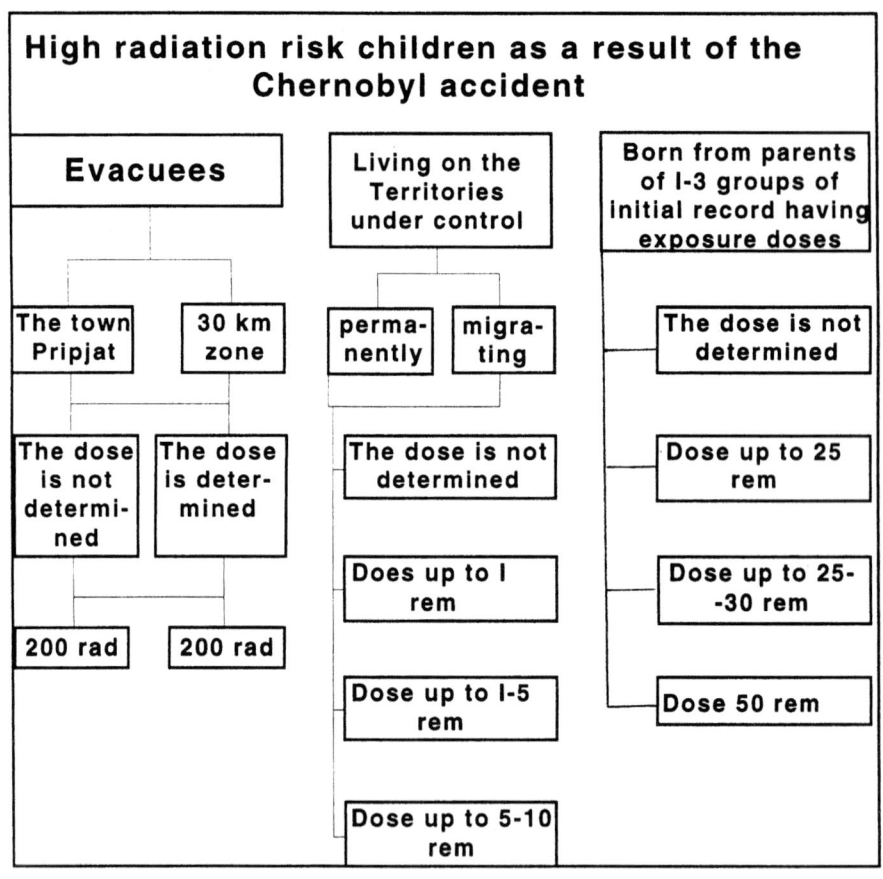

controlled areas. Average data on contamination for these areas is given in table 1. For most of them a "spotted" structure of Cs-137 contamination is typical, by virtue of which its density in different settlements of one territory may differ tens or even hundreds times. Thus, minimum contamination density in the Narodichi district is 0.5 Cu km^2, maximum - 59 Cu km^2, in the Polessky district - 0.2 and 91.3 Cu km^2, in the Luginsky district - 1.0 and 12.3 Cu km^2, in Ovruch - 0.5 and 11.7 Cu km^2, in the Ivankovsky district - 0.3 and 4.2 Cu km^2, respectively. It's obvious that children living in these territories need a long clinical surveillance as they are one of the high risk groups (see the diagram).

Morbidity is an indicator of the state of the environment [5,6]. Morbidity of children up to 14 years of age was studied from 1986 to 1988 according to the frequency of referring to the doctor and results of medical examinations carried out annually within the framework of clinical examination of the child population.

As a result, some peculiarities in the structure, levels and dynamics of general morbidity and separate nosologic forms have been detected.

Table 2. The dynamics of children's morbidity (per 10,000 of children population) of the 0 to 14 year age group, living on the territories under control in the Ukraine

Territory	All diseases		
	1986	1987	1988
The Zhitomir region including districts	16196.1±20.4	15550.0±18.9	16306.0±20.6
Luginsky	12218.2±162.6	13243.3±201.5	17297.3±350.0
Narodichi	8805.2±48.0	12434.0±82.2	13367.0±99.4
Ovruch	10469.0±17.4	12051.3±39.0	13335.2±52.4
The Kiev region including districts	14918.0±17.1	15159.0±17.7	15846.1±19.2
Ivankovsky	11023.9±86.0	13373.5±79.1	14129.4±92.6
Polessky	9709.9±52.9	12480.4±63.1	13496.1±88.0
The Ukraine	15296.0±14.1	14543.1±13.3	15718.0±13.8

It is determined that, overall, morbidity in the period under analysis increased 1.5 times in the Narodichi district; in Ovruch - 1.25 times, in the

Ivankovsky district - 1.28 times, and in the Polessky district - 1.39 times (table 2). Its level in all areas being studies, except for Luginsky district, was lower in 1988 than in the Ukraine as a whole or in the corresponding regions. At the same time, statistical data given evidence of the large rate of growth of morbidity in the controlled areas as against the territories unexposed to radioactive contamination. Respiratory organ diseases occupy a leading place in the overall structure of children's morbidity (25.0 to 70 percent at different periods of observation). Acute respiratory virus infections amount to about 70 percent of respiratory organ diseases (Figure 1). Diseases of adenoids and tonsils are considerably prevailing.

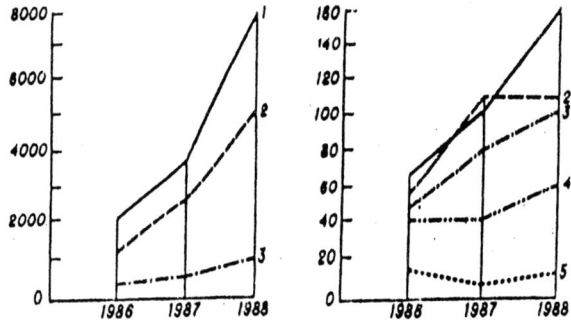

Figure1. The dynamics of morbidity from respiratory organs in children living on the territories under control of the Ukraine.
X- axis - years of observation; Y - axis number of diseases per 10000 people; 1 - diseases of respiratory organs; 2 - acute respiratory virus infections; 3 - chronic diseases of tonsils.
Figure 2 The dynamics of moridity from iron deficiency in children of the territories under control of the Ukraine.
x - axis - years of observation, Y - axis - number of diseases per 10000 people. 1 - I Ivankovaky region, 2 - Polessky region, 3 - Narodichy region, 4 - Ovruch region, 5 - Luginsky region.

A growth in pathology of the upper respiratory tract is noted in all the areas under observation, especially in 1988. Indicators of morbidity were in the Luginsky district 1041±95.4%, in the Narodochi district - 830.2±40.8%, in the Ovruch district 499.9±17.1%, in the Ivankovsky district - 554.4±27.7%, and in the Polessky district - 494.7±25.6%. These values are much higher than the level of morbidity with long-term tonsil disease and adenoid in children living in other areas of the Zhitomir and Kiev regions (355.0±9.7 and 390.0±3.8%, respectively). An increase in the incidence of morbidity from pneumonia was distinctly pronounced only in the Ivankovsky and Ovruch districts, in 1988 in particular (547.1±27.6 and 324.5±13.9%, respectively). In 1986, these figures were 409.0±24.0 and 162.0±10.6%, respectively. Pneumonia morbidity changed little over the whole period of observation and is consistently higher in the Polessky district. In the Ukraine this index was 256.0±0.6% in 1986 and 249.0±0.6% in 1988.

Other respiratory organ disease, (chronic bronchitis, bronchial asthma), were insignificant but had a tendency to increase. Diseases of the digestive organs have a considerable specific weight in the structure of overall morbidity. Morbidity from this pathology from 1986 to 1988 is characterized by a phased nature of changes, however its growth is noted in practically all areas. High cholecystitis morbidity was recorded in 1988 in the Ivankovsky district (407.4±23.9%) and the Polessky district (146.3±15.3%). These diseases are 12.8 and 1.1 percent of the total structure of morbidity. It should be noted that the level of cholecystitis morbidity was higher in districts of the Kiev region as compared with other territories in 1986 to 1987. In 1988, the index of morbidity in the Ukraine amounted to 45.6±0.3%.

The specific weight of infections and parasitic diseases in the structure of overall morbidity in 1986 was 1.8 percent on the average per district; in 1987 it was 2.0 percent and in 1988 - 2.6 percent. A reliable increase in the incidence of this pathology [p 0.05] during a 3 year period of observation took place in the Ivankovsky and Polessky districts, its morbidity in 1988 reached 205.8±17.2 and 437.2±266.1%, respectively. Intestinal infections were predominant in this group. The authors have failed to reveal regularities in the dynamics in children of the Narodichi district. At the same time, a more low lever of morbidity was recorded in 1988 than in previous years 0 529.3±33.1%. However, this value characterized a maximum level of child morbidity of the districts under control.

Morbidity from disease of blood and hemopoietic organs in children living on the territories under control increased, first of all, due to iron deficiency. During the 3 years after the accident the incidence of iron deficiency increased 2.6 times in the Ivankovsky district; 1.9 times in the Norodichi district; 1.8 times in the Polessky district; 1.6 times in the Ovruch district (Figure 2). Indicators of morbidity from iron deficiency in the Ukraine from 1986 to 1988 were increasing insignificantly.

Mental disorders occupy a definite place in the structure of child morbidity on the territories under control. It should be noted that their incidence in the areas in question did not exceed corresponding figures in the Ukraine as a whole for the first 2 years of observation after the accident. At the same time, an increase in the incidence of mental disorders on the territories affected by radioactive contamination was noted, (an average level of growth for the 3 years is 26.0 percent). Thus, in 1988, morbidity in the Ukraine as a whole was 317.5±0.7%, whereas in the Ovruch district it reached 417.2±15.7%, in the Ivankovsky district

297.0±20.5%, in the Narodichi district 281.1±24.4%, in the Luginsky district 250.1±31.2%.

Morbidity from nervous system disease and sense organs was almost the same in children of the territories under observation from 1986 to 1987 and averaged approximately 150.0±6.5%. In 1988 an increase in the given pathology was noted. Chronic otitis is about 10 percent of all pathology of the nervous system, the specific weight of children's cerebral paralysis is smaller. The data obtained gives evidence of an increase in morbidity from children's cerebral paralysis. Maximum morbidity was recorded in the Narodichi district in 1988 (30.3±8.1%).

From 1986 to 1988 the specific weight of congenital anomalies changed considerably. The highest level was recorded in the Narodichi district in 1988 - 1.7 per 100 deliveries (in the Ukraine - 1.9). During the period in question the highest level of congenital anomalies of heart and blood circulation system was preserved in the Luginsky district. The highest increment of morbidity is observed in this district, and in 1988 it made 103.3±31.5%. The incidence of the given pathology in the Narodichi and Ovruch districts is half as low as in Luginsky district, and the increment of morbidity is inconsiderable. Even lower figures are typical of children of the Ivankovsky and Polessky districts - 27.9±6.3 and 34.5±7.5%, respectively (1988).

Morbidity from malignant neoplasms among children of the areas under control agrees with the average values recorded in these areas. Thus, patients with a diagnosis made for the first time in 1986 were recorded only in the Polessky district (3.4%, in the Ivankovsky district - 1.3% in 1988 (in the Ukraine - 1.68%). No cases of malignant neoplasms were detected in children of the Narodichi district from 1986 to 1988.

The data obtained gives us every reason to believe that the Chernobyl accident has caused definite changes in the health status of children living in the territories affected by radioactive contamination.

First of all, changes in the total resistance of a child's body are observed. A recorded growth of morbidity of children of the territories under control outstrips the corresponding figure in other districts in the Kiev and Zhitomir regions and in the Ukraine as a whole. It should be noted that distinctions in figures of morbidity in 1986 and 1988 in all areas under observation are reliable [p 0.01]. A growth of respiratory diseases attracts one's attention. Sample surveys show that in separate children they are recorded up to 8 to 10 times a year without connection to seasonal variations. The incidence of disease of the upper respiratory tract is much higher in children involved in the emergency situation than in children living in areas unaffected by radioactive contamination. The dynamics of morbidity of digestive organs and cholecystitis, in particular, are also characterized by a growth in the districts under the control of the Kiev region.

It is important to note that a growth in iron deficiency is more frequent. The reasons for this are not unambiguous.

The whole set of natural and anthropogenic factors available in the contact area at a definite time has an influence on the health status. It requires a systematic consideration of the problem, application of modern methods of statistical analysis: correlation, factor and cluster analysis, whose underlying criteria is the dangers for people's health. A classifier includes diagnostic symptoms, and qualitative and quantitative parameters corresponding to a concrete type of situation.

The ecological situation is described by means of separate indices or indicators. A classifier represents a complex of special estimation scales which allows one to determine a degree of danger in the separate factors according to the values of separate indices as well as a degree of danger in a group of factors by a combination of values of the set of indices. Estimates of a situation would be different for regions with different natural conditions, anthropogenic influence upon the environment and with the different health status of the people.

The availability of information on a complex of parameters characterizing the quality of water, food, and health of the people is a prerequisite for the organization of studies on the analysis and assessment of factors of the environment and on elaboration of a classifier of ecological situations.

A computer-aided database "Ecology" (CDBE) that can solve this problem is being developed at the Institute of Epidemiology and Prevention of Radiation Injuries of Ukrainian Scientific Center of Radiation Medicine. It provides for the acquisition of initial data compatible with territory and time, computation of indices of radiation and nonradiation contaminations, and the selection of indices-indicators of external health factors. A data bank is being developed which contains the amount of contaminants released into the environment, the state of its components, (atmosphere, water reservoirs, ground water, soil), and the contamination of food and potable water.

To make CDBE the following materials are used: statistics of the national economy on the release of contaminants and the use of natural resources, (state land paperwork, intradepartmental paperwork of Agricultural Industry), paperwork of the State Committee of Hydrometeorology

TABLE 3

FACTORS OF ANTHROPOGENIC CONTAMINATION AND THEIR INDICATORS (ACCORDING TO THE MATERIALS FOR THE KIEV REGION)

Factor's rank	Factor	Index-indicators of factors	Correlation coefficient (K) with the dynamics of morbidity of children in the 1 to 14 year age range						
			total	iron deficiency	pneumonia	tonsilitis	cholecystitis	congenital anomalies of heart vessels	otitis
1	Anthropogenic radioactive contamination with Cs-137	Maximum Cs-137 contamination density in the administrative district	0.64	0.85					
2	Load of mineral fertilizers on the environment	Average speed of change of density of application of fertilizers							
		potash	0.56						
		Phosphate	0.53						
3	Load of pesticides on the environment	Average speed of change of density of application of pesticides:							
		herbicides			.73				
		mercury containing pesticides			.66				
4	Load of contamination on water reservoirs	Average speed of change of density of discharge of sewage into water reservoirs:							
		biologically purifies pure water according to specification				0.72			0.63
5	Load of atmospheric pollution	Average speed of change of density of sulphur dioxide release into the atmosphere						0.63	

on the results of the analysis for radioactive contamination of locality; paperwork of the epidemiologic service on the results of analysis for contamination of food and water, objects of the environment with residual amounts of pesticides and nitrates; data on the quality of water in water reservoirs at sites of water managements; data on the contamination of food and potable water with radionuclides; analysis made by the laboratories of the Ukraine Scientific Center of Radiation Medicine.

The analysis of indices of the ecological situation and children's morbidity has made it possible to rank factors of the ecological situation by the extent of their effect on people's health (table 3). In the Kiev region the following external factors have the greatest importance: anthropogenic radioactive contamination and the loading of mineral fertilizers and pesticides on the environment.

Further investigations will allow us to confirm a more reliable connection between the factors of the ecological situation with people's health in the territories affected by the accident. It will reveal a method for modelling the dynamics of health, depending on factors of the environment, which would promote the development of a classifier of ecological situations and methods for assessing the state of the environment according to the data of people's health and vice versa.

REFERENCES

1. *Biological effects at the prolonged intake of radionuclides.* - M., 1988.
2. **Iljin L.A., Balonov M.i., Buldakov L.A.** et al. *Med. radiology.* - 1989. - No 11. - pp. 59-81.
3. **Knizhnikov V.A., Barchudarov R.M., Bruk G.J.** et al. Medical aspects of the Chernobyl accident. - Kiev, 1988. - pp. 66-76.
4. **Moskalev G.I., Kidritsky J.K.** *Med. radiology.* - 1983. - No 4. - pp. 70-74.
5. **Sidorenko G.J., Mozhaev E.A.,** *Sanitary state of the environment and people's health.* - M, 1987.
6. **Shandala M.G., Zvenjatskovskaja L.I.** *The environment and people's health.* - Kiev, 1988.
7. **Goldman M.** *Science.* - 1987. - Vol. 238. - pp. 622-623.

MEDICAL CONSEQUENCES OF THE RADIATION ACCIDENT IN THE SOUTHERN URALS IN 1957

L.A. Buldakov, S.N. Demin, V.A. Kostyuchenko,
N.A. Koshurnikova, L.Yu. Krestinina, M.M.Saurov,
I.A. Ternovskij, Z.B. Tokarskaya and V.L. Shvedov

On September 29, 1957, because of a fault in the cooling system of concrete tanks containing highly active nitrate-acetate wastes, a chemical explosion occurred in the materials and radioactive fission products were released into the atmosphere and subsequently scattered and deposited in parts of the Chelyabinsk, Sverdlovsk and Tyumensk provinces. The aggregate amount of activity released amounted to about 2×10^6 Ci (7.4×10^{16} Bq). The composition of the material released is indicated in Table 1.

For the area with a ^{90}Sr contamination density of 0.1 Ci km^2, (double the level of global fallout), the maximum length of the deposition track under the radioactive plume form reached 300 km; for ^{90}Sr, contamination density of 2 Ci km^2, it reached 105 km, with a width of 8 to 9 km. The area of density distribution is shown in Table 2.

The presence of gamma emitters among the contaminating nuclides was responsible for the external irradiation of the population and the environment. During the initial period, the dose rate was about 150 R h in the area with a ^{90}Sr contamination density of 1 Ci km^2, with maximum values of 0.6 R h at the head end of the track where the contamination density (^{90}Sr) reached 4000 Ci km^2.

Because of radioactive decay of the short-lived nuclides, contamination levels and gamma dose rates in the area of the accident fell off fairly rapidly during the first few years after formation of the deposition track, (see Table 3), and subsequently the radiation situation was governed entirely by the presence of strontium-90 and its rate of radioactive decay.

Table 1. Characteristics of the Radionuclide Mixture Released in the Accident

Radionuclide	Contribution to total activity of the mixture, %	Half-life	Type of radiation emitted	Nature of radiological hazard
^{89}Sr	traces	51 d	β, γ	
$^{90}Sr + ^{90}Y$	5.4	28.6 y	β	Internal irradiation (skeleton)
$^{95}Zr + ^{95}Nb$	24.9	65 d	β, γ	External irradiation
$^{106}Ru + ^{106}Rh$	3.7	1 y	β, γ	External
^{137}Cs	0.036	30 y	β, γ	External and internal
$^{144}Ce + ^{144}Pr$	66	284 d	β, γ	External
^{147}Pm	traces	2.6 y	β, γ	
^{155}Eu	traces	5 y	β, γ	
^{239}Pu	traces	-	α	

Table 2. Area and Population of the Contaminated Region

Density of radioactive contamination, $Ci/Km^2 (^{90}Sr)$	Area, Km^2	Population ($\times 10^3$)
0.1	15 000	270
including:		
2	1 000	10
100	120	1.5
1000	20	1.154

Table 3. Dynamics of the Radiation Situation

Time after accident, years	Contamination density		Gamma dose rate (relative to initial value)
	Gross activity (relative units)	^{90}Sr (Ci/Km^2)	
0	1	0.027	1
1	0.34	0.026	5.6×10^{-2}
3	0.10	0.025	8.3×10^{-3}
10	0.043	0.021	9.8×10^{-4} due to
25	0.029	0.014	3.8×10^{-4} ^{137}Cs

The exposure of the population in the contaminated territory was due, in the first instance, to external irradiation from the soil and from objects in their surroundings - including their own clothing - and also to internal irradiation due to the consumption of contaminated food and drinking water and inhalation at the time when the cloud was being formed. Subsequently, (after half a year to a year), internal exposure from contaminated food was predominant.

The radiation protection measures adopted for the population were as follows:

- Evacuation of the population;
- Decontamination of some portions of the agricultural land;
- Monitoring of contamination levels in agricultural produce and rejection of produce with activity levels exceeding the accepted norms;
- Limitations imposed on the utilization of contaminated land;
- Reorganization of agriculture and forestry, with the creation of specialized state farms and forestry enterprises operating in accordance with the special recommendations worked out in the light of the accident.

The dynamics of the evacuation exercise for people living in regions with a ^{90}Sr contamination density above 2 Ci km^2 are shown in Table 4.

Table 4. Dynamics of Population Evacuation and of Exposure Dose to the Population before Evacuation

Population group and size ($\times 10^3$)	Average contamination density, $Ci/Km^2/(^{90}Sr)$	Time required for evacuation, days	Average dose received up to evacuation, rem	
			External exposure	Effective dose eq.
A: 1.15	500	7–10	17	52
B: 0.28	65	250	14	44
C: 2.0	18	250	3.9	12
D: 4.2	8.9	330	1.9	5.6
E: 3.1	3.3	670	0.68	2.3
Total: 10.83				

In the immediate aftermath of the accident - that is, within 7 to 10 days - six hundred people were evacuated from the settlements in the most severely affected area; and about ten thousand people were evacuated in the 18 months following the accident. Altogether 10,180 people were evacuated. The maximum average exposure doses preceding evacuation reached 17 rem in external exposure and 52 rem in effective dose equivalent (150 rem to the gastro-intestinal tract). These doses can be doubled in light of the non-uniformity of contamination density and the conditions in which the exposure occurred.

Table. 5. Frequency of Abnormalities of Arterial Blood Pressure and Systole Frequency (Pulse Rate) Among Exposed Individuals Belonging to Population Group A

Index	Fraction of group (%) affected	
Tachycardia (pulse > 90)	(5.5)	4–7.4
Bradycardia (pulse < 60)	(8.5)	0–14.1
Hypertension (blood pressure > 160/95 mmHg)	(3.3)	1.7–4.0
Borderline hypertension (blood pressure 140/90–159/94 mmHg)	(10)	7.5–14.5
Hypotension (blood pressure < 110/60 mmHg)	(16.4)	10.8–24.0

Mass medical surveillance of the inhabitants of the affected region was arranged a year after the accident. It included examinations by pediatricians and therapists, by neuropathologists and gynecologists, peripheral blood analysis, and determinations of body weight and height. Factors entailing a risk of cancer were assessed, as was the cardiovascular pathology, and the presence of harmful habits; and urine tests for albumin and sugar were carried out. Serum cholesterol levels were determined and otolaryngeal examinations were organized; and all the individuals studied were given ECGs.

The portion of the population most seriously affected by radiation, (Group A), was comparatively young: people 0 to 17 years of age accounted for 45%, 18 to 45 years of age 39% and above 50 only 16% of this group.

In the clinical studies carried out on this population, no cases of radiation sickness were noted. During the early period of the investigations, 21% of the cases examined showed no decrease in the peripheral blood leukocyte count. However, the peripheral blood indices showed, in adults, average values for thrombocytes, (236-280 x 10^9 L), leukocytes (7.2-7.5 x 10^9 L), and neutrophils (4.1-4.7 x 10^9 L), which were no different from those found in normal unexposed adults. The distribution function for these indices during the first examination period was the same, in terms of median values, for the irradiated group as for the references, but among exposed people a larger percentage showed two-sigma deviations from the average. Thus, the fraction of individuals with leukocyte counts above 9 x 10^9 L was 17-19%; with thrombocyte counts above 350 x 10^9 L it was 7-8%.

The reaction of the cardiovascular system was studied in all the individuals investigated on the basis of arterial blood pressure and systole frequency (pulse rate). The results among the essentially healthy members of Group A indicated no regular increase in the frequency with which these indices departed from a normal distribution, nor any deviation as a function of radiation dose.

The population studies indicated that up to 75% were, for all practical purposes, healthy individuals. Twenty-five percent of those investigated revealed general somatic problems of one kind or another. Among these, as can be seen from Table 6, more than half were suffering from cardiovascular disorders, and almost 30% had respiratory illnesses.

Table 6. Nature and Frequency of Disease in the Individuals Studied

Class of illness, nosological units	Fraction of individuals affected, %	
Parasitic infection, helminthiasis	0.6	
Nodular goitre and thyrotoxicosis	0.5	
Psychic disorders, neurasthenia	1.9	
Circulatory disorders:		
Rheumatic heart disease	1.8)	
Hypertension	2.5)	
Ischaemia	3.3)	13.7
Coronary and cerebral atherosclerosis	5.1)	
Varicose veins	1.0)	
Respiratory disorders:		
Acute nasopharyngitis	5.3)	
Bronchitis	2.3)	8.3
Emphysema	0.7)	

Thus, the examinations conducted on individuals living in settlements at the head of the deposition track showed no clinical evidence of radiation pathology. It is fair to assume that certain distortions in the distribution of blood indices are associated with the haematological reaction to irradiation observed in the early period: leukopenia, relative lymphoplasia and a leftward shift in the neutrophil formula.

Long after the accident, medical studies were performed on individuals who could be said to belong to the critical group: these were people whose exposure to radiation had occurred during the period of body formation and development and in whom the exposure levels had been highest (Groups A and B). Of these individuals, one third were, for all practical purposes, healthy. In the rest, careful investigation showed chronic tonsilitis, and in another 16% chronic [word omitted in original] and cervicitis). The frequency of osteochondrosis increased with the age of the individual. Three people had epilepsy associated with alcoholism and cranial trauma. The morbidity of the exposed individuals revealed no special characteristics by comparison with the control contingent. Peripheral blood indices were in the normal range. With increasing age the frequency of dystrophic changes in the ECG increased (classes 4, 5 and 9 in the Minnesota code). The frequency of ECGs in the zero class, (with no changes), was no less in the irradiated persons than in the controls.

Serum cholesterol concentrations (mmol L) were the same among the irradiated contingent as among the controls: under 29 years of age 4.78 ± 0.1; up to 39 years of age 5.25 ± 0.06; up to 49 years 5.41 ± 0.06; and above 50 years 5.69 ± 0.06.

Table 7. Extensive (%) and Intensive Indices of Mortality Due to Congenital Developmental Anomalies

Population group	Extensive indices	Intensive indices after 35 years	
		per 10^5 population	per 10^3 newborn infants
1 - 10 270 individuals	0.36	4.2	1.02
2 - 23 230 individuals	0.38	4.2	1.93
K - 21 537 individuals	0.67	7.4	2.66
Chelyabinsk province			
- 1965	0.53	3.6	2.3
- 1986	0.23	2.2	1.3

Certain disorders thought to be cancer risk factors were recorded no more frequently among irradiated persons than among the controls. Thus, in the 28-year-olds, chronic gastritis, endocervicitis and cervical erosion were encountered in 2.3, 11.1 and 11.1% of cases, respectively, whereas in the 50-year-olds they were found in 9, 20 and 0% of cases. One of the sensitive criteria for damage caused by ionizing radiation in infant mortality and intrauterine developmental anomalies. Over the 35 year period, 35 cases of death due to congenital anomalies have been found among the offspring of the population living on land covered by the radioactive deposition track. In the first group, consisting of 10,270 individuals living in areas with a ^{90}Sr concentration of 1-2 Ci km^2, there were 10 cases, and in the second group, consisting of 23,230 individuals living in an area with a ^{90}Sr density of 0.1-1Ci km^2 there were 25 cases. In the control group, consisting of 21,537 individuals living in an area with less than 0.1 Ci km^2 (^{90}Sr), there were 39 cases of death due to congenital anomalies. In the overall mortality structure it was found that mortality due to developmental defects accounts for 0.36 to 0.67% of all cases (see Table 7).

As can be seen from Table 7, the differences between the groups are statistically unreliable; nor did any significant differences emerge during the first two years following the accident. The structure of the lethal developmental defects diagnosed in these groups is shown in Table 8. It will be seen that the fraction of chromosome anomalies among the lethal defects is largest in the population belonging to Group 1. Developmental defects of the circulatory system are also more frequent in Groups 1 and 2 than in the control contingent, although these differences are not statistically significant.

Table 8. Lethal Developmental Defects in Different Population Groups, %

Type/site of anomaly	International classification number	Population groups		
		1	2	3
Chromosomes	758	28.6	6.7	20
Nervous system	740 – 742	14.3	6.7	32
Circulatory system	745 – 747	–	6.7	–
Digestive tract	750 – 751	–	–	4.0
Other organs and systems	748, 749, 756, 759	14.2	6.6	8.0
Totals		100	100	100

Table 9. Mortality of Newborn Infants with Congenital Developmental Defects between 1980 and 1987 (per 1000 Live Births)

In the entire affected zone	At the nearest settlement	in Chelyabinsk province	In Sverdlovsk
0.95±0.08	1.7±0.4	1.0±0.08	1.1±0.07

The figures in Table 9 demonstrate the absence of any statistically significant difference, although they are higher for the settlement closest to the radiation source. The reason for these higher figures still needs further clarification.

Highly instructive data were obtained from an analyses of infant mortality during the years following the accident (Table 10). As can be seen from this table, there is no appreciable difference in infant mortality between the three groups compared, even against the background of the rather high infant mortality prevailing during those years. Here, too, the cases of infant mortality are apparently not associated so much with the levels of radiation exposure as with inequalities in the medical treatment accorded to newborn infants.

Table 10. Mortality of Infants Aged Up to 1 Years (per 1000 Live Births)

Causes of death	Territory covered by deposition track	Control No. 1 on track boundary	Control No. 2 far from track boundary
All causes	27.7	31.4	38.6
Nutritional disorders	15.2 ± 2.8	12.3 ± 3	5 ± 1
Pneumonia	1.7 ± 1.0	3.1 ± 1.5	16.1 ± 1.8
Infectious diseases	1.6 ± 0.9	2.3 ± 1.3	3.0 ± 0.8
Disease of the newborn	8.7 ± 2.2	13.8 ± 3.2	14.5 ± 1.7

Table 11. Size of the Exposed Population and Average Doses Received

Group	Number of inhabitants	Duration of exposure	Average doses, cSv					
			External γ-irradiation	Internal irradiation				Eff. dose equiv.
				G.I. tract	Lungs	Red bone marrow	Bone surfaces	
1	1 054	10 d	17	150	2.7	0.5	0.7	52
2	10 720	30 y	0.4	2	0.2	3.8	5.2	2
3	23 230	30 y	0.1	0.7	0.1	0.7	1.0	0.4
K	21 537	30 y	-	-	-	-	-	-

The remote (long-term) effects of radiation exposure were studied in parallel among the irradiated population and the control contingent, and also in a zone where the effects of a nuclear facility might be expected to make themselves felt. In this way, more than 100,000 people were surveyed. Table 11 shows the effects of exposure to radiation on the most severely irradiated contingent. Among these persons, intensive mortality indices in Groups 1, 2, 3 and K were, respectively, 272, 2760, 6578 and 5873 cases, respectively, and the corresponding mortality coefficients were 9.5, 11.5, 11.0 and 10.9×10^{-3}. It will be seen that there are no differences as compared with the control contingent.

At the same time, age-related mortality indices show substantial deviations from the control contingent in individuals under the age of 4 and older than the age of 60. Nevertheless, it has proven impossible to find any link with the radiation dose. Thus, in Groups 1, 2, 3 and K the mortality coefficients for children aged up to one year were 91, 32, 63 and 52, for children between one and four years of age 13.7, 1.7, 5.0 and 3.3, and for individuals over 60 years of age, 39.2, 50.4, 43.1 and 46.9, respectively. In all the other age groups the mortality indices and coefficients reflected no differences between the groups and the control contingent.

A fact to be noted is that among the 272 individuals from Group 1 who died, cancer as a cause of death comes in third place, after heart disease, injuries and accidents. Another peculiarity is the predominance of mortality from infectious diseases over that due to respiratory disorders.

Table 12. Extensive (%) Intensive ($\times 10^5$) Indices of Mortality Due to Maligant Tumours

Population group	Number of cases	%	10^{-5}	Confidence intervals, 95%
1	25	11.7	115.9	75-165
2	376	13.6	157.4	142-174
3	775	11.8	129.2	120-142
K	707	12.0	131.9	122-142

Of particular interest is the analysis of mortality due to malignant tumors since these are the principal late manifestation of the effects of irradiation. The highest mortality indices - on the boundary of reliable statistical significance - were noted in individuals belonging to Group 2 (see Table 12). However, the groups selected for the purposes of this comparison were not large enough and hence it is impossible to conclude definitely that there is any appreciable difference between the magnitudes observed.

Table 13. Structure of Mortality Due to Maligant Neoplasms (per 100,000 of Population)

Principal tumour sites	International classification number	Irradiated individuals			
		1	2	3	K
Oesophagus	150	26.5	8.2	11.3	12.1
Stomach	151	35.3	45.1	32.4	44.3
Other organs of the digestive system	152-159	8.8	30.7	20.9	22.4
Respiratory organs	160-163	17.7	29.5	24.9	26.4
Bones	170	0	3.1	0.9	2.4
Skin, oral cavity	140-147 172-173	0	7.5	1.4	4.5
Mammary gland	174	4.4	4.4	2.1	4.2
Corpus et cervix uteri	180-182	0	13.1	9.6	10.8
Other urogenital organs	183-189	4.6	9.4	6.6	7.6
Lymphatic and haematopoietic tissue	200-209	13.2	5.0	5.2	4.7

In the overall picture of neoplastic disorders observed, the most important place is occupied, (over the entire period of interest), by cancer of the digestive tract, and in particular by cancer of the esophagus (see Table 13). There is a tendency towards increased cancer of the esophagus in Population Group 1, which was the group subjected to the highest doses, but even so, these higher figures are not statistically significant.

Attention is also given to fatal cases of lymphatic neoplasms and neoplasms in haematopoietic tissue. The mortality coefficient in Group 1 was 13.2×10^{-5} as compared with 4.7×10^{-5} in other groups. Although these differences are not statistically reliable because they are based on only three fatalities, it should be noted that the EED in this group amounted to 52 cSv, which is close to the critical dose for leukemia induction.

Table 14. Mortality Due to Malignant Neoplasms

	In the whole of the affected zone	At the nearest settlement	In Cheyabinsk province	In Sverdlovsk province
1970-1980	145.8	-	146.6	-
1980-1987	160.7±2.5	105±12.7	167.6±3.2	159.4±6.6

The level of mortality due to neoplastic disorders, broken down by decades, is shown in Table 14 for the zone directly affected by the deposition track and compared with intensive indices for the neighboring regions Chelyabinsk (1) and Sverdolovsk (2). The figures show that lethality increases regularly from the first decade to the next. Overall, in the zone affected by the track and by the operations of nuclear facilities, it rose from 145.8 to 160.7 ± 25 per 100,000 of population, and in regions one and two, from 167.6 ± 3.2 and 159.4 ± 6.6 per 100,000. In the nearest settlement the figure was only 105 ± 12.7, although the radiation dose here was higher. This lower figure is due entirely to the younger average age of the population in this settlement.

Table 15. Percentage of Individuals who Married and had Children

Group	Age at time of accident	Number of individuals	% married	% with children
Infants	Under 1 year	56	91 (82-97)	84 (73-92)*
Children	1- 9 years	295	93 (89-96)*	90 (86-93)*
Juveniles	10-19 years	203	93 (89-96)*	93 (89-96)
Adults	20-29 years	201	95 (92-98)*	91 (87-94)
Adults	30-59 years	308	98 (96-99)*	98 (96-99)*
Controls (Russia as a whole)			81.9-82.6	94.6

* Significant differences from control.

Analysis of the causes of morbidity due to malignant neoplasms following the irradiation accident suggests a frequency grouping for the initially diagnosed tumors as a function of certain external factors. It was noted, for example, from the morbidity data in Chelyabinsk, that there was no connection between enhanced morbidity and dose rate. On the other hand, a clear and complete correlation was found between morbidity and releases of SO2 to the atmosphere. Although S02 is not itself a carcinogen, it is extremely useful as a gauge of general chemical contamination. Actual data show that when there are no SO2 releases morbidity amounts to 225 cases per 100,000 individuals per year, whereas in situations where SO2 is released in amounts of 50,000, 100,000 and 150,000 t per year the morbidity figures rise to 250, 275 and 300 cases per 100,000, respectively. Accordingly, on the Chelyabinsk map the cancer mortality figures correlate not with the radioactive contamination track but with the location of metallurgical and chemical plants.

Table 16. Dynamics of Birth-rate Coefficients Among the Evacuated Population (per 1000)

Time after accident (years)	1	5	10	15	20	25	30	1-30
Number of children	51	271	491	717	960	1242	1586	1616
Birth-rate coefficient	37.4	42.2	30.2	27.5	26.4	27.8	30.0	31.8
Standardized coefficients	40.4	48.7	31.8	26.9	24.8	26.2	26.6	31.8
Birth-rate coefficients for Chelyabinsk province	24.1	20.8	14.8	16.0	16.7	19.8	16.7	18.4

A great deal of attention has been given to the reproductive state of individuals irradiated at different ages. The figures in Table 15 indicate no systematic deviations from the norm, as far as this extremely important demographic indicator is concerned, among the individuals who received the largest doses. It will be seen from Table 15 that those who were newborn infants at the time of the accident and who married by the time they were 27 years of age have, as yet, comparatively few children. Among those who were older at the time, on the hand, marriage frequency has been higher than in the controls, and the number of children has been either no different or slightly less than those among the controls (this applies to individuals who were up to nine years old at the time of the accident). At the same time, the birth-rate coefficients per 1000 inhabitants, as seen from Table 16, are higher among the population in the affected region than in the district as a whole. One gets the impression that living conditions and social factors among the population evacuated at the time of the accident are somewhat more favorable than those among the rest of the agricultural population in the region. There may also be some other factors, such as special national characteristics involved.

In conclusion, it may noted that observations on health, morbidity and mortality among the population subjected to the accidental release of radiation - with whole body exposure doses from 1 to 52 cSv and irradiation of individual organs up to 150 cSv - have revealed no significant deviations from the comparable values found among healthy unexposed individuals.

CONSEQUENCES OF CHRONIC RADIATION EXPOSURE

L.A. Buldakov,
Academician of the Russian Academy of Medical Sciences

A.K. Guskova,
Corresponding Member of the Russian Academy of Medical Sciences
The Institute of Biophysics of the
Ministry of Public Health of Russia, Moscow, Russia

The nuclear industry created 50 years ago in different countries of the world was only connected with military potential at first. That is why, the medical aspects of the problem of radiation exposure in man were accessible for only a limited group of people for a long time. This fact is a reality for all countries with nuclear weapons.

However, an experience of the observation of workers of the nuclear industry and nuclear power and their family members covering many years and the surveillance in population living in near-by areas of nuclear installations and test sites now gives a real opportunity for reasonable estimates of different ecological situation consequences rather than making only the theoretical prognosis. This corresponds to problems that occurred in the regions contaminated after the accident at the Chernobyl Nuclear Power Plant (ChNPP).

This paper is a brief general review of some of the results received during the studies carried out for many years. The directions of the studies are:

- the study of early reactions and late consequences of radiation exposure, including the investigation of adults and children who lived in the area of the accident or in near-by regions with nuclear industry facilities;
- the preparation of the databank on the general and local radiation injuries which occurred due to radiation accidents (including the ChNPP accident);
- the treatment of the main clinical signs of acute radiation syndrome (ARS) and local radiation injuries (LRI) combined with the efficacy estimate for means used to modify the course of radiation syndrome;
- the improvement of the diagnosis and the expert estimation of the health condition in people who received single (acute) radiation exposure and those, who had prolonged exposure, (in a wide range of radiation doses).

One of the objects observed was the staff of the first nuclear facility in this country, the "Mayak" integrated facility. Unfavorable work conditions during the first 5-10 years caused chronical radiation syndrome (CRS) to occur. There were cases in 5.8 and 22.5 percent of the people who worked at two branches of this facility, (average accumulated individual doses of total radiation exposure were equal to 264+14 cGy and 340+5 cGy, respectively).

To reveal the effect of chronic exposure the threshold dose in critical organs should be more than 50-100 cGy received within a relatively short period of time, (the dose rate is more than 10 cSv year, but more frequent incidence and expression of the disease were observed for 25 cSv year).

The decrease of the exposing intensity, (less than 5-10 cSv year and a total accumulated dose greater than 1 Sv), is accompanied with a clear recover process, which practically eliminated the previously revealed changes within 3 to 5 years.

The confident improvements of the radiation exposure induction of incidence which cause an increase in such diseases as cardial ishemia, brain insult, hypertension, osteochondritis with secondary radix syndrome, atherosclerosis, ulcer disease, and gastritis were not found.

Nevertheless, some correlation between several groups of diseases of the lungs, kidneys, and gastrointestinal tract and typical occupational conditions were found. However, the expressiveness of this correlation is very weak and probably reflects the indirect integrated influence of the whole multitude of exogenetic and endogenetic factors, local conditions, diagnostic and treatment approaches of physicians observed. The accumulated data requires the immediate improvement of the quality of physician care in Medical Departments and Facilities and strict diagnosis in most widespread diseases influencing workability, which predominantly determine the incidence and the profile of mortality of the facility staff.

The confident relationship between the disease and radiation exposure in people engaged in the nuclear industry during the beginning of the industry operation was only revealed for oncological pathology and hemoblastosises. It is also marked in some population groups exposed by the radioactive wastes of the "Mayak" facility or those who lived in the area of the nuclear test site during the atmosphere nuclear tests. The detected excess of the incidence, (compared with the control group), as a rule, correlates with the exceeding determined dose level, (1-1.5 Sv in bone marrow, 2.5-3 Sv in bone tissue and lungs, and much higher levels in the liver and thyroid gland).

The recovering dynamics and outcomes of CRS were studied. 632 CRS patients were observed within 30 to 35 years, i.e. they were studied from the moment of the disease revealing and elimination of the patient contacts with radiation. Hemopoises recovering was established in 88.1% of all cases. In other cases, the moderate general or partial, (White blood growth and megacariocytes), hypoplasia of the hemopoises was indicated. For the doses of more than 2 Sv, an increase in the forming of cerebral atherosclerosis, (of the people less than 50 years of age), was found, which was expressed in a moderate change of the immune status, in the persistent increased, (compared with a spontaneous one), level of structure damages of the somatic cell genome (1 lymphocytes). The mortality structure of CRS patients keeps malignant tumors in the first place, particularly lung cancer in personnel internally contaminated with alpha emitting nuclides.

The standard of living and dose levels in two towns situated in the nearest area of the first nuclear facilities were also studied.

The change in the dose level and its partial ingredients corresponding to the gradual decrease of gas and aerosol waste (in 30 to 100 times), occurred within the observation period. Respectively, the critical organ doses have also decreased (for instance, in integrated facility No. 2 they reached the level of several percent of the annual maximum permissible dose (MPD); the total accumulated doses have decreased from 17.3 cSv (in the 50th) to 1.1 cSv (today).

The standardized indices of the mortality rate of the population lived near "Mayak" during 1950 to 1988, as well as the dynamics of the child mortality rate for each period of the 5 years, demonstrates the lower values in comparison with the average indices in Russia. The general mortality rate in 1950 to 54 was 630 and 940, in 1960 to 64 it was 320 and 720; it became equal to 675 and 1030 in 1980 to 88 (all values relate to 100,000 of the population in town No. 1 and to general population of Russia respectively). During these same years the child mortality rate (per 1000 births) was 51.8 and 75.2; 22.2 and 31.8; 14.3 and 25.9, respectively. One can see that the persistent radiation exposure on the level of 0.5 cSv year did not have a severe effect on the main demographic indices.

The analysis of the average annual indices of the population mortality rate for a period of 30 years, (for town No. 2), demonstrated the intensive mortality coefficients before and after standardization intended to be lower than the coefficients for the control group (regional administrative center): 128.6, (town No. 2), and 138.6, (in the regional center, per 100,000 people per
year. No peculiarities were found in the sexual-age mortality coefficients calculated for the same period.

Data on the dynamics of oncological mortality indicate only the relative increase, (in comparison with controls), for town No. 2 in 1982 to 1985: the standardized mortality coefficient, (the accepted age standard is a population of the regional center, in this period reached 156 cases, (town No. 2), and 148.9 cases in the regional center per 100,000 people per year. Comparing this data with the general values for this country demonstrated the standardized mortality coefficient for town No. 2 to be equal to 184.7 and the general coefficient is 145.1 (per 100,000 of people per # year). The confidentiality criterium differs from the control group (91 and 95 percents, respectively). Some relative increase in mortality for town No. 2 at this period is marked for lymphogranulomatosis and cancer of the respiratory organs.

The subject of interest consists in the analysis of some peculiarities of the clinical conditions related to the specificity of the radiation factor origin. The generalization of a number of investigations of the exposure of unsoluble compounds of alpha emitter (Pu-239), medical observation data, and single-directed complex surveillance in the staff who worked with this radionuclide was done. The study established the lung changes, (restrictive type of the outer breathing function (OBF) disturbance), and lymphopoises changes (the moderate lymphopenia correlating with radiation dose). Occupational pneumosclerosis was only established for these people who were engaged at the facility in the first start-up period and received 1.2-2.7 Gy of the internal lung exposure. Average annual dose equivalent at this very period exceeded the permissible occupational limit in, approximately, 10 times. The increased risk of lung cancer and

OBF decrease in smoking workers was found at the dose levels of one MPD. Lymphopoises changes, (lymphocytopenia, structure changes in lymphocyte genome), were found. The correlation between lymphocytopenia and cytogenetic effects and absorbed dose in tracheobronchial lymph nodes was established first. It is suggested that the mutagenetic effect, (increasing of the chromosome aberration counts for lymphocytes of the peripheral blood), and immune imbalance signs can have a significant place in the pathogenesis of tumoral effects for this radionuclide.

The same observations were done for the facility, where the tritium represents the main radiation harm. The analysis was done taking the contact duration and specifity of the work operations with trithium for different occupational groups into account.

If the CRS was caused by tritium, (in the late period of observation which covered 30 years), the majority of cases was remarked to have clinical recovery. However, as in the case of CRS caused by external radiation exposure, the increase of early cerebral atherosclerosis incidence was remarked.

At the late period, (10 to 25 years of observation), in the case of sufficient overexposure in the past, the (the total accumulated dose is 200 to 600 cSv per 4 to 5 years), for the small number of cases, (32 people), the moderately expressed hypoplasia of granulocytopoisis, the changes of cellular immunity, and persistent chromosome aberrations in peripheral blood lymphocytes were found.

For chronic internal exposure and internal exposure combined with external gamma exposure, (at the level of doses closed to MPD), the changes of health conditions, which can be correlated with occupational factors, (the observation period is 25 years), were not found.

Based on accumulated experience, the methodological recommendations on the optimization of the system of revealing, surveillance, prophylaxis, hospitalization terms in different diseases of people engaged with radiation sources, and risk group forming were developed.

The study of the health of two generations of people who worked at the oldest nuclear facility still continues. The estimate of the health of grandchildren, (1564 persons), up to the age of 3 was first done in comparison with the control group collected from the population of the same towns who did not have the occupational exposure.

These investigations are supplemented by information on the health conditions of children who were directly exposed by the radiation waste of "Mayak" at the first period of its operation, (1945-1957), and also by the information on the children of the facility personnel.

The dose levels could reach 100 cSv and *in utero* irradiation could be more than 5 cSv during first the 3 months of pregnancy.

Using clinical data, laboratory tests, demographic and genetic methods, the analysis of the posterity of health in the first, (16,000 people), and second, (about 2,500 people), generation from parents exposed by the nuclear fission products, (average individual gonade dose is 16 cSv with dose groups from 3 to 100 cSv), was done.

No differences were found in the conditions of the main organs and organism systems, in the incidence of the innate defects of development (IDD), and in the structural damages of lymphocyte chromosomes. However, the increased level of general morbidity is marked and this morbidity consists mainly in the respiratory tract diseases with a complicated course and frequent contagious infections is now trying to explain these factors.

The first database covers 2000 grandchildren of the staff and town population. Each child is described by 200 characteristics including health conditions and factors influencing different development stages.

According to the registry data, (16,000 children), the birth rate in the exposed population is not decreased and sexual relationship in born children is not disturbed.

The integral index of "the frequency of the hereditary pathology", which joins the information on the spontaneous abortions, lethal innate defects of development, Dawn disease, and early perinatal mortality was equal to 12.7% in the control group and 17% in the exposed population. This probably reflects the increased intensity of the mutagenesis process in the last group.

During the whole period of observation, (30 years), no mortality rate increase was found in infants, (up to 1 year old), of exposed parents. However, some change of infant mortality structure was marked: endogenious reason deaths are more frequent.

A comparative analysis of the 1st and 2nd generation did not reveal sufficient differences in such indices as death and incidence of the innate defects of development.

The above-mentioned materials are in close relation with the original data obtained on the estimate of the medical demographic situation in the population of the two most contaminated regions of Belorussia in the 5th year after the accident at ChNPP.

The levels of accumulated doses, (from 1986 to 1989), in average were equal to 14.5 cSv for the regional contamination density of 60 Ci km^2, these values were: 11.4 cSv for 30-60 Ci km^2, 6.4 cSv for 15-30 Ci km^2, 4.5 cSv for 5-515 Ci km^2, and 1.7 cSv for <5 Ci km^2.

The investigations were done in Gomel' and Mogilev regions for the general population, (according to the data of the State Committee on Statistics), and in separate groups of different exposure doses. (The indices were considered both for the whole population of each region and particularly for the rural population; the data were collected for 5 years before the accident and for 4 years after it.)

The results certify some decreases in the birth rate in the early period after the accident, (in comparison with the pre-accidental results). The highest change was found for the Gomel' region. If one accepts the average birth rate to be equal to the rate observed during 3 years before the accident (100%), in 1986, 1987, and 1988 it was equal to 136, 88, and 181 percents, respectively (in the dose range of 14.5 cSv); these values for different dose ranges are equal to: 60, 66, 22% (11.4 cSv); 92, 99, 116 % (6.4 cSv). Birth rate fluctuations are seen to be caused by a social factor rather than radiation exposure.

The general mortality rate in the population of the regions mentioned above and in Belorussia as a whole persists to be at a level close to the pre-accident one. The analysis of the data on the tumor mortality demonstrated its gradual increase from 1981 for both the Gomel' and Mogilev regions and for Belorussia as a whole, which is a widespread trend. No changes connected to the accident were found.

Infant mortality, (less than 1 year old), gradually decreased for all contingents. The mortality indices for the infections, parasitic diseases, and respiratory organ diseases reached the lowest level in the post-accidental period. The mortality for innate abnormalities of development in the children of the Gomel' and Mogilev regions was in the same pre-accident level and did not exceed the average Belorussian level.

The results of all the elaborated studies confirmed the absence of sufficient changes of medical demographic indices, which can be related to the early influence of the accident at ChNPP in the population of the Gomel' and Mogilev regions, (both in the total population and in the rural population), in 1986-89.

The dynamic observations of approximately 600 of the children living in the Krasnopol'e district of the Mogilev region, (a strict radiation regimen area), are preliminarily summarized for the 4 years of observation. Average absorbed individual doses for the thyroid gland were equal to 10 to 390 cGy; about 40% of the people investigated had a total accumulated radiation dose of 10 to 15 cGy and 60% had less than 10 cGy. Only in single cases did the probable doses reach 25 cGy.

According to the detailed clinical study data, significant changes in the somatic health condition relating to the accident are not revealed.

No increasing incidences of microcephalism, CNS development defects, and disemriological stigmas spectrum change in children born in 1986-89 (in comparison with those born in 1980-85).

Instability of the enzymic homeostasis characterized by multidirected changes of different hormone contents, (prolactine content increase, the increase of hormone content(for "stress group" of enzymes: cortisol and adrenocorticotropic hormone, the decrease of insulin content and increased gluconic content, ferritine increase), which was marked in 1986, was transient and this instability did not reveal a dose-dependent relationship.

In the following years the normalization of the majority of enzymic contents was observed and only the lower levels of testoid were found in boys and decreased ferritine levels were found for both sexes (less than 7 years old) in 1990. In general, one can conclude, that the children received an average individual thyroid dose of 260 cGy and did not give any significant changes of their enzymic homeostasis.

No dependence was found between the thyroid hyperplasm degree and radioiodine content and absorbed thyroid doses (dose range is 20 to 400 cGy).

Also, it was not established that the increase of the thyroid hyperplasm incidence in children depends upon the thyroid dose values. The negative influence of the antistrumine preparation administration was found to cause the change of incidence and severity of this pathology in the group of 80 cGy average individual thyroid dose.

The correlative increase of the incidence of other endocrine pathology, (obesity, diabetes, etc.), was found in the same children groups.

A significant incidence of iron deficiency anemias was found, particularly in small children. At the same time, the incidence of iron deficiency anemias in pregnant women living in the Krasnopol'e district of the Mogilev region in 1985 reached the values of 80%, and in other years it was equal to: 64.8% (1986), 56.4% (1987), 55.8% (1989). There was the simultaneous decrease of iron deficiency anemia incidence in newborns and in small children (less than 3 years old). It was observed on the levels of : 90% (1985), 79% (1986), 69% (1987), and 52% (1989).

The positive dynamics of the general anti-infection resistance in children was established in 1986-1990. However, in the age group less than 3 years old, the decreased level of this index continues to persist.

The analysis of the limited population group, (10,000 people), lived near the Semipalatinsk test site during the first intensive tests of the nuclear installations in the air and ground surface, (until 1958), demonstrated the results described below. The accumulated doses for a number of tests could reach 0.2 to 0.3 to 1.6 Gy in several settlements. Taking the spontaneous incidence of tumors into account in this contingent, (within 40 years it was 824 cases), the amount of tumors related to the exposure according to the accepted risk coefficients has to be equal to 7%, (55 cases), and for Semipalatinsk itself, less than 0.15%. The cases of CRS and ARS, the increase of thyroid pathology incidence, (i.e. the effects directly determined by radiation), were not revealed for this group of people. The sample cytogenetic studies of 100 people with the highest radiation doses demonstrated the increased chromosome aberration dissemination in this group in comparison with chromatid aberrations observed in controls.

The underground nuclear tests had an insufficiently low level of population doses and there was no influence on people's health in these cases.

Systematic clinical physiological studies and laboratory tests, which were elaborated on for many years, conclude, that in the red bone marrow dose range of 50 to 100 cGy year some adaptation reactions are revealed as well as an increases in the incidence of some late effects of radiation (leukemia, tumors); these adaptation reactions are statistically confident in population groups but they have no diagnostic significance in separate individuals.

In the population exposed by uranium fission products separate the shifts of the population distribution for the blood counts and for the immune indices in variation curves were found during the first years. They had a transient character and they were smoothed, when the dose rate decreased to 10 to 12 cGy year level.

The regular functions of the neural system have normalized more slowly; vegetative instability persisted, as well as the asthetic manifestations and microfocal symptoms. However, these deviations should not be considered as the direct consequences of radiation exposure, because of the absence of a strict correlation between these signs and the radiation doses, (which makes them different from some subclinical structures - bone and bone marrow - keeping such strict dependence as the later terms).

Today, the study of the health of personnel working with chronic radiation exposure generated by low doses of external and combined, (external exposure + internal intake of radionuclides), radiation exposure is continued. The programs revealing the increased risk groups, (for cerebral atherosclerosis complications and hypertension), are developing on the base of brain insult registry. The grandchildren's health continues to be studied 6 to 8 years of age for grandparents who worked in the nuclear industry.

Based upon the clinical departments of the Institute of Biophysics and its Branches consulting and health experts are doing different things for different groups of people. Together with local Medical Departments, the estimates of general and occupational workability losses are expanding to reimburse the health damages and to find optimal work conditions for future employment of the patients. The problems of adequate response to the population for "living with the radiation", which became an integral part of everyday life in this country, are now evolving.

Future improvement of the databank and the registry of ARS patients, which is focused in the Institute of Clinics, already gives one the opportunity to accept the information on a number of aspects of this disease.

In conclusion the authors would like to formulate some important statements summarizing the studies.

1. The main effects caused by radiation exposure, (the manifestations of ARS and CRS, the worsening of health indices of people due to the change of resistance for widespread harmful factors caused by people living and activities in the accidentally contaminated regions), occur only after exceeding the determined level of radiation exposure (usually, it is more than 25 cSv year and 100 cSv of total accumulated dose).
2. The changes, found for the doses marked above, have, in general, reversible features; after the elimination or decrease of the radiation exposure (less than 0.1 to 0.06 Sv year) the practical clinical recovery takes place.
3. The stochastic late effects of the polyethiological character in people exposed to radiation can be revealed statistically confident as the difference between their incidences and background incidences; it is possible only for doses, which usually exceed the permissible occupational limits now accepted (0.05 Sv year and 1 Sv of the total accumulated dose).

THE EXPERIENCE OF THE FIRST NUCLEAR FACILITY (EXPOSURE LEVELS AND STAFF HEALTH)

B.V. Nikipelov[1], A.F. Lyzlov[2], and N.A. Koshurnikova[3]

[1] Deputy Minister of the Ministry of Atomic Power Industry of Russia
[2] Head of Radiation Protection Facility
[3] Head, Member of Scientific Staff of the Branch
Institute of Biophysics of the Ministry of Public Health of Russia

In August 1949, the first Soviet A-Bomb was tested at the Semipalatinsk Test Site. This explosion put an end to the United States' nuclear weapons monopoly and averted the real threat of nuclear attack, which had existed for almost 5 years prior to this test.

The Soviet Atomic Project was headed by I.V. Kurchatov, and the work was carried out by thousands of Soviet scientists, engineers, and laborers, including those who worked at defense Facility A and Facility B situated in the Cheliabinsk region.

Facility A consisted of a uranium channel nuclear reactor with graphite moderation of thermal neutrons and direct water flow cooling. It was started up in June, 1948; it has since ceased operating and has been dismantled.

Facility B was a radiochemical production plant used to separate uranium and plutonium from spent nuclear fuel. It was started up in December 1948; it has since been dismantled.

The primary harmful occupational factor in both facilities is ionizing radiation. During the first years of the operation of these facilities, gamma radiation was the most significant form of exposure. This has been confirmed not only by measurement of gamma exposure with individual dosimeters, but also by the appearance of health problems normally considered to be the result of gamma radiation.

From the first days of nuclear research, the problems of radiation protection were constantly under consideration. In August 1948, the Ministry of the Atomic Power Industry (MAPI) and the Ministry of Public Health of the USSR prepared "General Sanitary Standards and Health Protection Regulations for People Employed at A and B Facilities." This document was approved separately at each facility. According to the dominant concept of dose tolerance, the permissible daily (6 working hours per day) dose established for staff was equal to 0.1 cSv (this is equivalent to approximately 30 cSv per year). In the case of an accident, a singleexposure of 25 cSv over 15 min was permitted, and a medical examination after such exposure was obligatory. When such exposure occurred, workers were required to take a leave of absence with pay or temporarily work in an area where they would have no radiation exposure.

Dosage monitoring was performed at each facility from the beginning of their operation. The industrial production of dosimeters was begun at that time in order to allow the measurement of gamma radiation exposure. Individuals wore film badge dosimeters, which had an energy range of 0.4–3 MeV and a dose range of 0.05–3 cSv (30% error). If the dosimeters were read after each shift a workers, turned in their badges for the day, then such a method provided reliable estimates of daily doses. The individual dose was measured at the location of the badge (chest pocket or trousers pocket of overalls), and this dose was considered to be the total body radiation dose.

A special system of sanitary and medical surveillance was created for nuclear industry. A.I. Burnasjan was the first organizer and was the head of this system for several decades. It is necessary to recognize the significant contributions made by G.D. Baisogolov and A.K. Guskova to the health care of nuclear industry staff; they are distinguished persons considered to be the founders of the school of radiation medicine in this country. The joint efforts of engineers and physicians improved the occupational radiation protection system. The dangers of high occupational radiation exposure were understood not only by supervisors, but also by workers. Workers also had the same degree of understanding of the necessity of nuclear weapons for this country despite the health risks of such weapons.

OCCUPATIONAL EXPOSURE LEVELS AND MEASURES TO DECREASE THE LEVELS

The nuclear reactor and the radiochemical plant were designed to keep all dangerous processes under reliable biological shielding and control.

Nevertheless, equipment malfunctions, technological communications disturbances, and malfunctions of monitoring devices occurred not only during the start-up period, but also during the first experiments, typical for new technology. These malfunctions resulted in frequent radiation overexposure of the staff.

As experience was gained, technology was improved, means of protecting individuals were upgraded, and staff exposure doses were decreased. The past 40 years can be divided into several periods which are distinguished by rather different occupational radiation exposure levels (see Figure 1). In 1948–1952, protection processes had just been developed, and the radiation situation was heavily studied in order to decrease occupational exposure. Measures to decrease radiation exposure were begun to be actively used in 1953–1959. This provided the opportunity to normalize the radiation situation, and resulted in international standards forradiation exposure, which were developed in 1960–1973.

The data in Table 1 show that the percentage of Facility A workers who received doses of greater than 25 cSv significantly decreased in 1950–52. Facility B, which was started-up one half-year later, had a much different building design, this is why almost half of its staff received more than 100 cSv per year in 1950–51.

In Facility A, most of the radiation exposure occurred in the central hall, the storage pools, and during product transportation. However, during the same periods, staff members situated in a number of other rooms were barely exposed. During the early days of radiation studies, detailed maps of radiation fields throughout the reactors were made; these formed the basis for the determination of the amount of time one could work in various rooms. The scheme of equipment placement in Facility B caused radioactive contamination in practically allrooms. This is why a significantly higher percentage of Facility B staff received greater radiation doses than did Facility A staff.

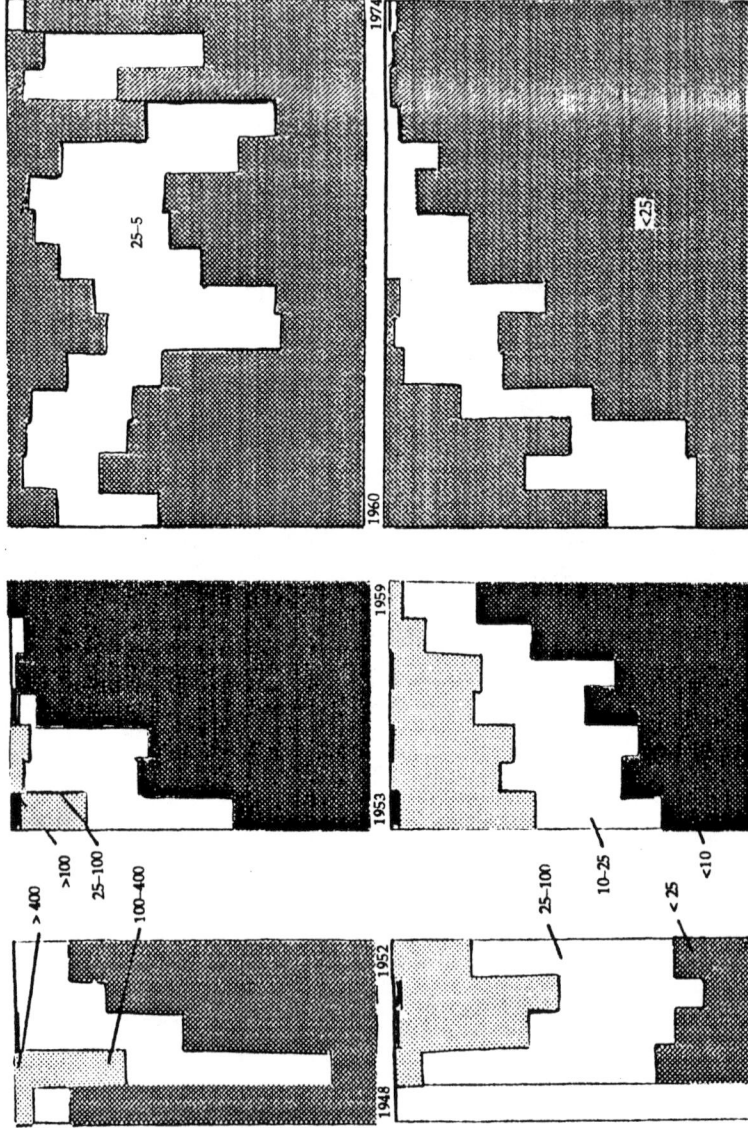

Figure 1. The time dynamics of dose distribution in staff (Upper distribution—Facility A, lower distribution—Facility B). Values are the dose intervals in cSv.

Table 1
Staff Radiation Dose Distributions (% of people).

Year	Facility A Doses, cSv					Average dose, cSv	Facility B Doses, cSv					Average dose, cSv
	<25	25-100	100-400	<400			<25	25-100	100-400	>400		
1948	84.1	11.1	4.8	-		19.6	-	-	-	-		-
1949	10.7	57.7	31.1	0.5		93.6	26.9	66.2	6.9	-		48.0
1950	52.2	47.2	0.6	-		30.7	21.5	42.0	36.0	0.5		94.0
1951	74.9	25.1	-	-		18.1	13.8	41.6	42.8	1.8		113.3
1952	83.9	16.1	-	-		14.9	21.8	57.0	21.2	-		66.0
	<10	10-25	25-100	>100			<10	10-25	25-100	>100		
1953	37.8	41.5	18.4	2.3		19.6	25.3	25.4	47.3	2.0		30.7
1954	64.0	33.0	3.0	-		8.9	34.7	36.1	29.1	0.1		20.0
1955	61.8	33.7	4.5	-		9.5	29.8	36.7	33.3	0.3		21.3
1956	92.3	6.4	0.6	0.7		5.1	45.0	31.9	23.1	-		16.2
1957	98.1	1.9	-	-		4.2	37.5	36.9	25.5	0.1		17.5
1958	93.5	4.7	-	-		4.4	59.6	31.3	9.1	-		10.8
1959	99.7	0.3	-	-		3.3	75.7	21.1	3.2	-		14.7

Table 1 (Continued)

	<2.5	2.5-5	>5		<2.5	2.5-5	>5	
1960	57.5	29.3	13.2	2.7	14.2	25.8	60.0	15.2
1961	73.9	22.4	3.7	2.0	13.8	49.1	37.1	11.0
1962	65.0	31.4	4.0	2.3	16.6	32.5	50.9	7.6
1963	64.3	29.8	5.9	2.4	41.4	37.3	21.3	3.8
1964	55.7	27.8	16.5	3.0	66.4	29.3	4.3	4.1
1965	24.5	49.1	26.4	4.0	67.0	31.4	1.6	2.1
1966	25.5	52.4	22.1	1.7	56.7	41.3	2.0	2.4
1967	45.5	41.4	13.1	1.3	76.7	23.2	0.1	1.8
1968	55.0	38.7	6.3	1.1	76.3	23.7	-	1.8
1969	56.2	39.5	4.3	1.0	91.9	8.1	-	1.4
1970	36.9	49.7	13.4	1.4	85.6	14.4	-	1.6
1971	25.7	36.5	37.8	1.3	95.1	4.9	-	1.4
1972	69.7	26.8	3.5	1.1	97.9	2.1	-	1.3
1973	45.5	44.8	9.7	1.0	97.4	3.6	-	1.3
1974	95.1	4.9	-	1.0	98.9	1.1	-	0.57

Naturally, the high dose exposure resulted in adverse health effects. The first cases of radiation were detected in early 1949. Minutes of a MAPI meeting held in 1949 describes the MAPI leaders' discussion of a letter written by A.P. Egorov (hematologist) to L.P. Beria (Minister of State Security); this letter deals with occupational radiation exposure in Facility A and Facility B. The "underestimation of occupational radiation exposure among facility supervisors, and some surprises inthe real radiation situation" are recorded in these minutes. It was decided that measures to improve working conditions be taken.

The degree of adherence to standards of occupational exposure and the degree of preparedness of new specialists for a very complex industry are difficult to assess. Even the project leaders (I.V. Kurchatov, A.P. Alexandrov and E.P. Slavsky) received significant doses of radiation. Nevertheless, administrators and medical service personnel in their attempts to decrease staff exposure. The "Statements on the Monitoring of Worker Health," introduced in 1949, required supervisors to immediately report to physicians cases of exposure at levels greater than 10 permissible daily limits (> 1 cSv). The administrative limitation system was created to prevent workers from performing tasks involving high radiation exposure without the written permission of supervisors.

In March 1950, the Collegium of MAPI decided to:

reconsider the staffing and the structure of dosimetrical services in Facilities A and B;

carry out weekly analysis of individual exposure data and to take immediate measures to decrease radioactive contaminations of the workplace;

punish those who violate exposure standards without just cause.

Weekly reports on occupational exposure were introduced.

One important factor that influenced the radiation situation, and consequently, the occupational exposure levels, was the system of organizational and technical radiation countermeasures. This system included a thorough inventory of radiation point sources (the "foci of radioactive contamination) in rooms. The elimination of these foci required a high level of organization, preparedness, and highly qualified staff.

Due to administrative, organizational, and technical countermeasures, the radiation situation in Facility A and Facility B was brought under control by the end of 1951. However, facility repairs were accompanied by significant radiation exposure. The number of repairs needed did not decrease, and the size of the repair staff was insufficient. Therefore, research staff also helped with repairs, resulting in a gradual decrease of their individual doses. In 1952–53, a system to limit exposure during repair procedures was completed; it was called the "access system." Each request for repairs involving high radiation exposure was approved in a written form ("access") which contained a description of the workplace, the permitted duration of work, and a list of protective measures to be followed.

So the period from 1948 to 1952 is characterized by high radiation doses and by the active efforts of administrators and medical personnel to bring radiation exposure to within permissible levels.

The new dose standards introduced in 1952 set the permissible exposure dose to 0.05 cSv during 6 working hours (15 cSv per year). New regulations established a single maximum permissible accidental exposure dose of 25 cSv for 15 minutes. These standards, which were acceptable for research staff, could not be maintained during repair procedures. In order to decrease the length of time needed for repairs (and consequently, the degree of radiation exposure), the most experienced and qualified workers were employed. However, due to the limited number of such people, it was not possible to effect a decrease in their individual doses. As a result, the regulation requiring those who were exposed to a dose of 100 cSv or more to then works in areas without occupational radiation exposure was adopted.

Details concerning this regulation were established by the MAPI and Facility Instructions adopted in 1954–55; 6 months of work in radiation-free conditions followed doses exceeding 45 cSv for the whole year, or a combined dose exceeding 75 cSv for the past two years.

Due to countermeasures introduced in 1953, the average dose from occupational exposure in Facility B was 2.15 times less than that in 1952. The percentage of persons who were exposed to doses greater than 25 cSv decreased from 78.2% (1952) to 49.79% (1953). Similar results were observed in Facility A in 1954. Some lag in the decrease of dose levels in Facility A staff in this year and later years can be explained by the fact that there always were a lot of complex repairs made in the reactor, which increased the radiation hazard. Despite this, the level of occupational exposure for the reactor staff was significantly lower than that for the staff of the radiochemical plant. This may be observed in a comparison of the average individual doses and the percentage of people, with the highest degree of exposure.

The document "Sanitary Regulations for Works With Radioactive Substances" was introduced in 1960. Supplement 2 of this document established the maximum permissible dose (MPD) of external exposure per week at 0.1 cSv and the annual MPD at 5 cSv. However, this document also established that the accumulated annual MPD could be increased to 12 cSv for people older than 30 years (Section 8 of Supplement 2). In 1970,

MAPI accepted the "standards of Radiation Safety" (SRS-69), which set the annual dose at less than 5 cSv (this level exists in currently used Standards (SRS-76 87).

Since 1968 (in Facility B), and 1974 (in Facility A), the annual individual doses of staff have been less than 5 cSv and the average levels of exposure continually decrease. So, due to the directed joint efforts of the administration and medical personnel, it became possible to decrease occupational exposure to a permissible level.

OVEREXPOSURE AND OCCUPATIONAL HEALTH

The Occupational Health is a reliable criterion used to characterize working conditions for any industrial facility. As was noted, the early years of Facility A and Facility B were characterized by external gamma radiation exposure being the primary harmful occupational factor. The incidence of radiation sickness can then be considered an early health effect of gamma exposure; mortality from malignant growths can be considered a late health effect.

Table 2. The Incidence of Chronic Radiation Syndrome (CRS) and Patient Radiation Doses

Facility	CRS incidence (% of entire staff)	Average gamma radiation dose, cSv	
		Accumulated Throughout Duration of Employment	Maximum Annual Dose
A	5.8 + 0.5	264 + 14	127 + 11
B	22.5 + 0.6	340 + 5	150 + 4

In the early 80s, the authors prepared registers including the entire staff of Facility A and Facility B. (Register results were obtained in cooperation with N.S. Shkol'nikova. The fate of 87% of Facility A staff and 85% of Facility B staff is now known. In order to analyze the registers, the records of those who began working in 1948 to 1958 were separated from the remainder of the records. This was done for two reasons. First, these people have worked under most harmful conditions (i.e., they received critical doses of external gamma exposure), and second, 30–40 years after the beginning of their contact with ionizing radiation sources had elapsed. Although this is not a long enough period for the full effects of radiation to be realized, it does cover the latent period for the majority of malignant growths.

Medical records (including archival materials were used to estimate the level of occupational morbidity. All cases of radiation sickness identified in Facility A and Facility B staff by the medical facilities of the Third General Department of the Ministry of Public Health of the USSR were considered.

Both reactor staff and radiochemical plant workers were chronically exposed to ionizing radiation. According to the dynamics of dose accumulation, chronic radiation syndrome (CRS) appeared more frequently in the reactor staff. Cases of acute radiation syndrome were high in both facilities, but were more frequent in the reactor staff, where a greater number of accidents, and consequent single "acute" exposures occurred (essentially during the initial period of reactor operation). It is necessary to stress that there were no fatal cases of ARS in the research repair staff of Facility A and Facility B.

Data analysis demonstrates that the CRS morbidity level and the individual external gamma doses in Facility B personnel were higher than in Facility A staff (95% confidence in Student's test). These differences are in close correlation with the differences in exposure in the staff of each of thesefacilities. Actually, the percentage of those who worked at Facility B in 1950–52 and had radiation doses exceeding 100 cSv year was 2–3 times greater than the percentage for Facility A.

The dynamics of dose formation is in close correlation not only with the CRS morbidity level, but also with the time of CRS appearances. Reactor staff received maximum radiation doses in 1949 and in that same year, radiation disease were first diagnosed. Maximum external gamma exposure in Facility B staff was recorded in 1950–52, and chronic radiation syndrome in these people was observed on year later than CRS appeared in Facility A staff.

Despite the differences in the time of appearance of CRS, CRS has only developed in those who had significant overexposure (see Table 2). The average accumulated dose of external gamma exposure in CRS patients is equal to 300 cSv and the maximum annual dose exceeds the standard that was established at that time (4–5 times more than 30 cSv year) and, of course, exceeds the modern limit (30 times more than 5 cSv year).

Thus, the clear dependence of the CRS morbidity level and its dynamics certifies that the appearance of CRS in both facilities was caused by external gamma exposure. The dose values, which are specific for development of this disease, indicate the existence of a relatively high threshold for the development of the disease.

As the radiation situation improved, the CRS morbidity level decreased. Thus, the percentage of CRS patients who suffered in 1950–53 is equal to 80%, but in 1954, it is equal to 7% (all percentages are calculated for the entire number of CRS patients). CRS did not appear in any who worked at the reactor after 1953, or in any who worked at the radiochemical plant after 1958.

Table 3. Oncological Mortality in Facility A and Facility B Staff (based upon the number of people who started working before 1958) Staff

Facility	Percentage of those who died from cancer who had an accumulated gamma radiation dose of:		Percentage of those who died from cancer who had a maximum annual dose of:	
	< 100 cSv	> 100 cSv	< 25 cSv	>25 cSv
A	5.7 + 0.6	9.4 + 1.2	5.9 + 0.7	8.7 + 1.1
B	4.3 + 0.4	8.1 + 0.6	4.2 + 0.5	7.7 + 0.5
Both	4.8 + 0.4	8.4 + 0.5	4.9 + 0.4	7.9 + 0.5

Standardized aged data are presented for the index of mortality in those who started their work during the same time period (i.e., the mortality index of contemporaries is given here). A statistically reliable increase in mortality in high dosage groups can be correlated with radiation dose. The mortality level in lower dosage groups does not differ from the oncological mortality level in the general adult population of the USSR (approx. 200 cases per 100,000 people per year, which is equal to approx. 6% for a 30 year period).

CRS patients, as well as a majority of the people overexposed during the first years of the nuclear industry, are still monitored by the specialized medi-cal facilities of the Third General Department of the Russian Ministry of Public Health. Moreover, an epidemiological survey has been conducted to study the late-appearing consequences of radiation exposure in these people. Data on malignant tumor mortality levels (including growths of hemopoietic and lymphatic tissues) of the Facility A and Facility B staff who began working in 1948–58 are now being collected (see Table 3). The average oncological mortality level is equal to 6.5% in the staff of both facilities. An expected dependence of tumor appearance on radiation dose is revealed; the mortality level in this case was determined by both the accumulated dose and the maximum individual dose per year. A reliable increase of mortality can be seen with accumulated doses greater than 100 cSv and with a maximum annual dose equal to 25 cSv for any year.

The oncological mortality for each facility does not vary greatly even though the radiochemical facility staff had contact with unsealed radionuclides, including transuranium alpha-emitters (plutonium, essentially), which for reactor personnel did not necessarily have contact with. The minimal difference in mortality despite the internal irradiation of Facility B staff may also be explained by the fact that only a relatively small percent of the staff worked with an amount of plutonium that could cause exposure comparable to external gamma irradiation.

The focus of this study, which is the effects of gamma radiation, does not mean that the authors deem internal exposure to be of little importance. The authors are certain (and this conviction is based not only on literature data, but also on personally-conducted experiments) that internal radionuclide exposure (alpha-emitters, essentially) has a significantly higher influence on the generation of malignant tumors than does external gamma exposure.

The data reviewed here came to form the basis for the formation of an at-risk group composed of people exposed to external gamma radiation at an accumulated dose of 100 cSv and a maximum annual dose of greater than 25 cSv. This group was identified in the Recommendations of the National Commission on Radiation Protection. The supervisors of MAPI and the Ministry of Public Health of the USSR received financial support from the Council of Ministers of the USSR so that payments could be made to those who was damaged, including those belonging to the at-risk group.

CONCLUSION

Due to demands for rapid production of nuclear weapons materials, employees of the first nuclear facilities in the USSR were subjected to high radiation exposure levels. This occurred because of organizational defects, imperfect technology, and a lack of experience, among other reasons. The result was an increase in occupational morbidity (however, at that time, morbidity did not exceed the level in industries requiring hard labor).

Nuclear facilities are very complex and their operation requires accurate organization, highly qualified staff, and firm discipline of workers; constant attention must be given to personnel safety. If all the necessary conditions are adhered to, then development can occur without harm to facility staff and the general population. This is confirmed by the experience of the first nuclear facilities, due to continual improvements of the radiation protection system, working conditions significantly improved in the late 50s, and exposure levels no longer exceeded permissible limits from early 70s onward. Not oneperson who began working at the reactor or the radiochemical facility after 1958 had developed CRS.

The investigations which were carried out in these facilities over a period of 40 years provided an opportunity to estimate penetrating radiation doses, which are a real health hazard. Not only acute radiation syndrome, but also chronic radiation syndrome can appear when radiation doses are much higher than permissible levels.

The lessons learned during the past 40 years of experience with nuclear materials are of great value to today's nuclear power industry.

CHANGES IN REPRODUCTIVE FUNCTIONS OF MICE INDUCED BY ISOLATED AND COMBINED ACTION OF X-RAY RADIATION AND STRESS

V.L. Vaskan
Institute of Radiation Hygiene, St. Petersburg

The problem concerning the response of an organism to the combined impact of radiation and stress factors of various origins is gaining importance in modern theoretical and practical medicine. The modifying influence that emotional strain-induced stress can exert on the impact of the low doses of ionizing radiation (IR) is believed to be crucial for the results of epidemiologic and clinical physiological analyses of various categories of the population exposed to additional irradiation s a result of the accident at the Chernobyl NPP.

It is noted in the ICRS document N 37 [3], that beside irradiation-inflicted health damage manifesting itself, for instance, in remote effects, it is necessary to consider other potentially harmful factors, such as awareness of risk and anxiety caused by the existence of dose limits. possible adverse consequences of the latter are hard to quantify.

Existing scientific data concerning the consequences that combined action of IR at low doses and emotional stress has for a human body are clearly insufficient in assessing their impact on socially significant health characteristics, in general, and reproductive functioning, in particular.

This study is aimed at experimentally assaying the effect that combining low doses of IR and stress has on the reproductive function.

Material and Technique

The research was conducted on mongrel white mice, (358 females and 120 males), aged 2.5-3 months, obtained from the "Rappolovo" breeding laboratory. The animals were mated for 5 days. In order to offset the influence of males reproductive characteristics on the outcome of the experiment, every male was mated with three females belonging to different groups. All the animals in the experiment were divided into following groups:

1. Physiological control (PC) - the animals were kept under usual conditions and have been subjected to no controlled influences.
2. Sham irradiation (SI) - on the 6th day after the mating, the animals, while in their permanent cages, were carried into an x-ray laboratory where they were subjected to manipulations, (without actual radiation), for 3 minutes and then brought back.
3-6. Groups subjected to the isolated influence of x-ray irradiation, with doses of 5, 25, 50, and 75 rem, respectively, animals comprising these groups underwent the same manipulations as those of SI groups but with added radiation exposure.
7. Immobilization stress group (IS) - the animals comprising this group underwent a daily 6-fold one-hour immobilization from 6th to 11th day since the mating started. The animals were held in individual cells with ventilation holes; they maintained a physiological pose but were unable to move or turn.
8-11. Groups subjected to a combined impact of radiation at doses of 5, 25, 50, and 75 rem, respectively, and daily 6-fold immobilization stress - animals of this group were subjected to the same manipulations as those of SI group, (but with radiation exposure added), and those of the IS group.

Radiation was performed on the RUM-17 apparatus on the 6th day since the mating started (E_{ef}=83 keV; U_A-200kV; anode current - 15 mA; CFR - 70 sm; filter - 0.5 mm Cu + 1 mm Al; dose rate - 26 R min.).

All the females were sacrificed by cervical dislocation on the 18th day since the mating started. A number of viable and dead fetuses present in the uterine horns were counted together with that of the yellow bodies in the ovaries. The following characteristics were used to describe the state of the reproductive function (RF); fertility rate, pre-implantation death (PreI), postimplantation death (PostD), and overall interuterine death (OID).

The results of the research have been statistically treated by means of the x^2 variance criterion technique used to compare alternative distributions: the exact Fisher method; the dispersion analysis of one-factor complexes for qualitative characteristics; the dispersion analysis of two-factor non-uniform complexes and the technique aimed at verifying the differences between two regression series.

Experimental data obtained are present in the table.

Characteristics of mice reproductive functions for isolated, combined and action of irradiation and stress

Group	Irradiation dose, rem	Mated females number	RF characteristics, %			
			F	PreD	PostD	OID
1.	PC	49	63.6	8.4	5.4	13.4
2.	SI	26	61.5	15.7	11.0	25.0
3-6. Isolated irradiation	5	30	63.6	8.9	9.2	17.4
	25	29	55.2	17.9	7.9	24.5
	50	29	68.9	25.2	10.7	33.2
	75	32	65.6	24.6	13.5	34.8
7.	IS	50	60.0	12.7	11.2	22.5
8-11. Combined action of irradiation and stress	5	30	60.0	28.1	6.1	32.4
	25	30	43.3	33.7	8.8	39.6
	50	29	41.3	35.4	21.3	49.2
	75	27	33.3	38.4	15.3	47.8

RESEARCH RESULTS

Fertility rate. The results of the data analysis have not displayed any certain differences between fertility rates of SI and IS group.s, or that of the PC group. Fertility rate of the groups subjected to isolated influence of irradiation does not vary with a dose from either the control of from SI.

The combined impact of irradiation and stress causes fertility to decrease with increasing irradiation dose. Strength of the radiation dose influences amounts of up to 15.0+2.3% of all the factors ($F>F_{001}$).

Fertility changes induced by a combined action of radiation and stress go at a lower rate than those induced by the isolate irradiation ($F>F_{01}$). Results of dispersion analysis show that in the case of a combined action of irradiation and stress, the fertility rate is affected by the stress factor alone ($\eta=3.5\%$ when $F>F_{01}$). No particular influence of the radiation factor has been detected on this material ($\eta=1.0\%$).

Preimplantation death. It has been detected that in the case of a SI the PreD rate increased in comparison to its PC value; with the increase similar to the one induced by immobilization stress. When the pregnant females received a 5 rem dose, their PreD rate decreased in comparison to SI, ($p<p_{05}$ according to EFM), and was similar to that of PC; when the radiation doses were equal to 50 and 75 rem, this characteristic was significantly higher than in PC and SI groups ($x^2 > x^2_{05}$). This conclusion is supported by the results of the one-factor dispersion analysis - $\eta^2=2.4+0.4\%$ when $F>F_{001}$.

The highest levels of PreD have been observed in the case of combining irradiation and stress - 28.1-38.4%. PreD level resulting from a combined impact of irradiation and stress is higher than the case of isolated radiation for all of the administered doses or in the case of the isolated influence of stress ($x^2 > x^2_{03}$). Two-factor dispersion analysis suggests the certain influence of both stress factor, (=2.3 when $F>>F_{001}$) and the radiation dose factor $\eta^2=1.2$ when $F>F$).

Postimplantation death. The PostD rate of the SI group is twice that for the PC group ($x^2 > x^2_{05}$) and does not differ from the Post?D level induced by a more prolonged and serious stress caused by a 6-fold one-hour immobilization. No certain differences between PostD rates for all administered doses and that of the SI have been detected on the material available, while the PostD rate of the animals exposed to the 75 rem radiation dose appeared to be higher than that of the 25 rem dose ($p<p_{05}$).

Dose dependence of the PostD rate variations has been observed in the case of a combined action of radiation and stress. The most dramatic growth of PostD is detected for radiation doses equal to 50 and 75 rem. The strength of the radiation dose factor can influence amounts up to 3.5+0.6% of all the factors ($F>F_{001}$). Though the combined influence of irradiation and stress does not differ from the isolated irradiation influence judging by the mean level of PostD, the criterion of the unparallelity of the processes is equal to 2.7 ($F>F_{05}$). When stress is combined with a radiation dose of 50 rem the PostD rate is higher than in the case of an isolated irradiation at the same dose ($x^2 > x^2_{05}$). The results of the tw-factor dispersion analysis suggests that PostD is influenced by the radiation dose factor $\eta=1.1\%$ when $F>F_{0.1}$. The strength of the stress factor influence amounts to only 0.2%.

Overall interuterine death. SI and IS groups are characterized by the same OID level which is almost twice that of PC, ($x^2 > x^2_{01}$).

In the case of an isolated radiation dose of 5 rem, the OID rate does not differ significantly from that of the PC group. With the dose increasing up to 50 to 75 rem, the OID undergoes a certain increase, compared to the control group. The strength of the irradiation dose influence amounts of up to 2.1+0.4% in this case ($F>F_{001}$).

When irradiation and stress are combined, the dose dependence of OID is also observed. In this case, the strength or irradiation dose influence amounts to 2.0+0.4% of all the factors ($F>F_{01}$). It should be noted that a combined action of irradiation and stress results in a certain increase of OID, compared to that of the isolated radiation case for all of the administered doses ($x^2 < x^2_{05}$). Though the changes in OID in the case of combined action of radiation and stress have the same direction as the ones induced by isolated radiation, the former are more dramatic ($F>>F_{001}$). The results of the two-factor dispersion analysis prove to have certain influences over both the stress factor ($\eta=2.1\%$ when $F>>F_{001}$) and the irradiation dose factor ($\eta-1.9\%$ when $F>>F_{001}$) on the OID.

DISCUSSION

It is common knowledge that when a normal progress of pregnancy is disrupted, the interuterine development will be determined less by the origin of the irritant than the irritant's activity time. The sooner into the gestation period that the irritant is activated, the more profound the changes it induces in the development embryo will be. Early stages of interuterine development are more vulnerable to detrimental influences [4,5].

The author's data concerning the impact of isolated radiation on the reproductive function agrees with the literature which presents proof of an increase in the interuterine death rate when mice are exposed to radiation at doses 5 to 25 R in the early stages of interuterine development [7,8].

A high sensitivity of an antenatally development organism to the influence of immobilization stress affecting the maternal organism has been detected. Motion restriction, (temporal immobilization), is a considerably strong emotional irritant [6].

An analysis of the impact of a combined action of radiation and stress on the PF state has shown that stress irritations enhanced the post-radiation effect, namely, they resulted in a fertility rate decrease and growth of embryo's interuterine death rate.

Fertility reduction induced by the combined action of irradiation and stress is considered to be a result of the preimplantation death of embryos in a fraction of females.

An increase in the interuterine death rate in the immobilization of animals, much more pronounced in the case of a combined action of irradiation and stress, leads us to believe that those changes are closely linked to the regulation of gestation processes ensuring pregnancy preservation, full term of pregnancy and vegetation support. Similar results had been obtained earlier in the works analyzing the influence of stress irritants on the organism [1,2,4].

Results allow us to state that the so-called "isolated" radiation is, actually, a result of the combined action of radiation and stress caused by various manipulations that the animals had to undergo; the latter are practically impossible to eliminate completely.

Thus, it has been shown that the influence of emotional strain-induced stress, practically neglected before, in assessing possible health implications for those exposed to radiation, is a significant factor in itself, capable of affecting the health to an extent superior to that of the ionizing radiation. Stress factors are also able to dramatically modify the post-radiation effect.

REFERENCES

1. **Arshavsky I.A.** *Physiological mechanisms and trends of individual development.* - Moscow: Nauka,1982.- 270 p.
2. **Bal'magiya T.A., Surovtseva Z.F.** Specifics of rabbit fetal growth in the case of normally progressing pregnancy and in the case of gestation dominant retardation *Bul. exp. biologii i meditsiny.* - 1974. V. 38 - N4. - p. 44-47.
3. *Radiation protection optimization based on the cost-profit ration analysis:* Publ. N37 ICRS - Moscow: Energoatomizdat, 1985, 96 p.
4. **Savchenko Yu., Kovalevsky V.A.** Influence of the irritation of emotional zones of rat hypothalamus on the pregnancy progress and progeny development *Zhurn. vyshey nervnoy deyatelnosti.* - 1985 - N6. - p. 1104-1109.
5. **Surovtseva Z.F.** On the characteristics of growth specifics and fetal development in rabbits in the case of an experimentally disrupted pregnancy. - *Briefs of candidate of medicine thesis.* - Moscow, 1969.
6. **Mason J.W.** Emation as reflected in patterns of endocrine integration *Emation: their parameters and Measurement.* - New York: Raven Press, 1975.
7. **Ohzu E., Makino S.** Some abnormalities produced by low dose x-irradiation in early mouse embryo *Pric. Jap. Acad.* - 1964 - vol. 40, p. 670-673.
8. **Regh R.** Low levels of x-irradiation and the early mammalian embryo.*Am. J. Roentgenol. Rad. Then. Nucl. Med.* - 1962 - vol. 87, p. 559-566.

ONCOLOGIC MORBIDITY AND MORTALITY IN AREAS UNDER THE CONTROL OF THE BRJANSK PROVINCE FOLLOWING THE CHERNOBYL NUCLEAR POWER PLANT ACCIDENT

A.A. Dudarev, G.I. Miretsky, P.V. Ramzaev,
M.N. Troitskaya, and I.E. Shuvalov
Institute of Radiation Hygiene, St. Petersburg

The results of investigations on the prevalence of oncologic pathology for six districts of the Brjansk province, namely, Gordeev, Zlynka, Klimovo, Klintsy, Krasnogorsk, Novozybkov, (with a total 265,000 people as of 1986), affected after the Chernobyl accident, are given in this paper. The total population of the Brjansk province has been used as a control.

MATERIAL AND METHODS

The materials have been developed on the basis of information presented by the Directorate of Statistics of the Brjansk province, its Oncologic Clinic, (record card file), as well as data of the Institute of Radiation Hygiene, (St. Petersburg), on the situation in areas under control. The data collected on oncologic morbidity and mortality was conducted separately for each of the Soviets, (village, settlement, city), which allowed the dynamics of the factors in question to be analyzed independently of administrative zoning in the territory under control.

Table 1. Oncologic Morbidity and Mortality (General and Separate Nosologic Forms) of the Population in Areas under Control and in Brjansk Province in 1966-1990, M ± m

Annual Average per 100.000 Persons Standardized in Sex, Age and City/Vell of the Population in Areas under Control and in Brjansk Province in 1966-1990

Nosologic Form	Mortality		% *	Morbidity				City/Vell 1986-1990
	1981-1985	1986-1990		1966-1970	1971-1975	1976-1980	1981-1985	
1	2	3	4	5	6	7	8	9
General	160±7 (159±3)**	182±6 (172±6)	114 (108)	170±7	215±6	218±9	223±4 (266±10)	284±10 (315±10)
Oral Cavity	3,4±0,9 (3,8±0,2)	5,6±1,6 (5,5±0,4)	165 (148)	10,0±1,7	8,5±1,0	9,0±1,4	9,5±1,4 (11,4±0,6)	13,0±0,6 (14,5±0,5)
Esophagus	3,9±0,3 (4,6±0,3)	5,5±0,2 (5,0±0,3)	139 (108)	3,6±0,3	4,2±1,0	4,8±1,0	4,3±0,4 (4,8±0,4)	5,1±1,1 (6,3±0,5)
Stomach	47,8±1,6 (53,2±1,8)	45,8±2,9 (49,6±0,8)	96 (93)	41,7±7,9	66,9±4,6	61,8±3,5	54,5±1,3 (61,6±1,9)	50,9±2,4 (63,5±2,5)
Intestines	13,6±0,9 (15,5±0,4)	17,4±0,4 (16,9±1,1)	128 (109)	9,4±1,7	14,6±1,8	14,7±1,8	17,8±2,9	21,7±2,0 (13,6±1,3)
Liver	5,5±1,1	5,7±0,6	104	1,9±0,6	3,2±0,2	4,4±0,3	2,2±1,0	6,1±0,3
Pancreas	4,8±0,5	4,5±0,7	93	2,1±0,6	2,8±0,4	3,8±0,3	2,1±1,0	6,9±1,3
Larynx	3,3±0,8 (4,2±0,4)	4,1±0,3 (5,1±0,3)	125 (120)	2,6±0,8	4,0±0,6	4,3±0,4	3,3±0,7 (5,9±0,6)	6,9±1,1 (8,2±0,4)
Trachea, Bronchi, Lungs	30,1±1,1 (34,6±1,5)	29,7±2,8 (38,5±1,6)	99 (111)	25,0±3,0	23,1±1,9	28,1±2,1	31,3±2,5 (42,6±2,4)	37,0±2,6 (50,5±1,3)
Kidneys	2,2±0,8	3,2±0,4	150	1,3±0,4	2,6±0,6	1,4±0,5	3,7±0,9	3,5±0,6
Bladder	3,6±0,8	5,2±0,9	144	6,3±1,2	6,3±0,5	4,5±0,8	6,5±0,7	9,51±1,09 (9,81±2,18)
Kidneys+Bladder	2,9±1,0 (5,6±0,5)	4,2±0,6 (7,3±0,8)	146 (131)					
Mamma	7,7±0,9 (14,9±0,5)	15,4±2,6 (17,2±1,2)	199	14,4±2,0	17,6±1,7	23,6±3,7	23,7±2,5 (19,4±3,9)	32,6±2,7 (34,8±0,7)

Table 1 (Continued)

Uterus	3.3±0.5 (6.3±0.3)	7.7±2.5 (6.6±0.4)	237 (103)	4.9±1.3	8.1±1.1	14.7±0.8	12.3±1.2	17.9±2.2 (15.5±1.2)
Cervix of Uterus	5.9±1.0 (7.0±0.3)	8.7±2.3 (6.4±0.4)	148 (92)	22.1±1.2	21.5±2.7	18.2±1.9	16.2±1.7 (13.8±2.1)	17.5±4.1 (16.3±1.3)
Ovaries	6.3±1.3	9.8±1.8	156	6.3±1.2	9.5±1.4	10.0±0.8	10.3±1.5	13.4±2.1
Prostata	4.5±0.7 (5.6±0.8)	5.9±0.8 (5.4±0.5)	131 (97)	3.2±0.4	2.2±0.8	3.9±1.2	2.5±0.9	6.3±0.9 (8.9±0.5)
Skin	1.8±0.3 (1.9±0.3)	3.6±0.8 (1.9±0.2)	198 (98)	21.3±1.6	25.8±1.5	28.2±2.2	26.1±1.6 (28.9±1.3)	32.1±2.2 (31.3±1.6)
Bone Tissue	1.6±0.3 (2.2±0.3)	1.1±0.2 (2.0±0.1)	73 (91)	0.6±0.2	0.7±0.2	0.1±0.1	1.8±0.4	1.5±0.4 (4.0±0.02)
CNS	3.3±0.4	2.2±0.4	66	1.8±0.3	2.1±0.4	2.5±0.4	3.3±0.5	2.4±0.3
Thyroid	0.8±0.2	1.1±0.2	140	0.2±0.1	1.3±0.4	1.4±0.6	2.2±0.7	7.0±1.2 (5.3)**
Leukemia	3.45±0.65 (3.84±0.37)	4.06±0.32 (5.54±0.49)	118 (144)	1.52±0.51	2.21±0.66	0.04±0.04	2.28±0.84	4.25±0.3*
Lymphoma, Lymphosarcoma	1.0±0.2	1.9±0.6	185	0.5±0.4	0.5±0.4	0.4±0.2	1.2±0.5	2.9±0.9
Blood+Lymph	6.0±0.8 (8.4±0.5)	7.5±0.7 (9.5±0.5)	125 (113)	5.3±1.0	5.4±1.0	1.6±0.3	5.4±1.6 (9.4±0.4)	10.3±1.2 (12.2±0.8)

* – ratio of annual average indicator of 1986-1990 to annual average indicator of 1981-1985 in percentage;

** – indicators for the province as a whole are given in parentheses;

*** – indicator is calculated only for 1988 and 1989;
  – indicator is calculated only for 1988-1990;
  – indicator is calculated only for 1988;
  – including leukemia, myeloma, lymphogranuloma, lymphoma, lymphosarcoma.

Table 2. General Oncologic Morbidity and Mortality of the Urban and Rural Population in Areas under Control and the Brjansk Province from 1966 to 1990, M ± m

Annual Average per 100.000 Persons, Sex and Age Standardized

	Mortality			Morbidity					
	1981-1985	1986-1990	%*	1966-1970	1971-1975	1976-1980	1981-1985	1986-1990	%
City	162±6 (180±2)**	178±9 (180±4)	110 (100)	184±8	223±11	217±14	219±6	286±12	130
Village	154±8 (178±3)	191±12 (205±6)	124 (115)	141±9	198±12	220±6	230±7	282±7	123

*, ** - see notes to Table 1.

Table 3. General Oncologic Morbidity and Mortality of Urban and Rural Population in Areas under Control and the Brjansk Province in Six Age Groups from 1966 to 1990, M ± m

Annual Average per 100 000 Persons

	Age, Years	Mortality			Morbidity					
		1981-1985	1986-1990	%*	1966-1970	1971-1975	1976-1980	1981-1985	1986-1990	%
	1	2	3	4	5	6	7	8	9	10
C	0-14	6,5±1 (7,1±0,8)	6,7±1,1 (7,0±0,3)	103 (98)	6,3±2,4	5,5±1,2	6,6±1,7	8,8±1,1	15,5±3,8	176
	15-29	8,5±2,6 (11,1±0,7)	11,3±2,9 (10,8±0,8)	133 (97)	11,4±1,8	17,7±2,4	13,0±2,2	17,2±3,3	23,8±3,2	138
I	30-39	27,0±4,2 (29,1±0,2)	33,5±6,4 (31,5±2,8)	124 (108)	75,4±4,8	53,6±5,9	49,5±13,1	48,1±11,1	88,5±11,5	184
T	40-49	122±17 (144±7)	132±16 (142±4)	107 (99)	144±13	219±11	229±21	237±10	261±44	110
Y	50-59	371±18 (399±8)	309±7 (385±16)	83 (97)	422±34	428±49	503±58	541±18	569±43	105
	60 and	632±21 (707±13)	782±57 (738±25)	123 (104)	654 ±43	865±29	789±50	762±19	1095±59	144
V I L L A G E	0-14	8,9±1,2 (6,7±0,9)	7,8±2,7 (7,2±1,1)	88 (107)	3,5±0,7	6,6±1,6	2,8±0,4	8,3±2,2	12,2±3,0	146
	15-29	15,9±4,9 (13,3±2,2)	16,0±3,8 (13,9±1,4)	101 (104)	106±3,6	23,6±8,5	22,9±5,2	17,2±2,9	25,8±6,1	150
	30-39	25,9±6,6 (41,5±4,7)	28,9±10,5 (37,8±3,5)	111 (91)	47,9±10,2	48,7±7,9	42,1±9,2	52,4±7,4	84,9±15,3	162
V I L L A	40-49	96±11 (124±8)	111±9 (170±10)	121 (137)	94±13	169±18	203±24	162±15	189±17	116
	50-59	273±12 (341±13)	282±11 (362±16)	103 (106)	233±14	283±15	385±30	425±15	459±37	108

Table 3 (Continued)

		319+29 (347+10)	458+32 (433+27)	143 (125)	288+25	436+34	447+21	471+19	619+31	131
G E	60 and									
T O T A L	0-14	7,7+0,7 (6,9+0,4)	7,2+1,7 (7,0+0,4)	93 (101)	4,4+1,0	6,1+1,5	4,5+0,8	8,7+1,3	14,3+3,3	165
	15-29	11,8+1,1 (12,2+0,7)	12,9+1,9 (11,2+0,9)	11) (91)	11,1+1,0	20,3+4,3	17,0+3,4	17,1+2,1	24,5+3,5	143
	30-39	26,4+1,9 (33,9+3,3)	31,9+7,7 (34,3+3,5)	121 (101)	64,3+4,9	60,6+4,9	50,3+7,9	54,7+8,6 (67,8+4,8)	94,5+8,4 (77,9+1,4)	173 (114)
	40-49	110+14 (136+6)	125+11 (150+4)	113 (111)	113+ 12	190+15	215+22	200+10 (236+13)	230+16 (274+7)	115 (116)
	50-59	316+11 (376+4)	295+8 (377+10)	93 (100)	302+10	341+23	437+35	477+13 (551+16)	510+39 (612+20)	107 (111)
	60 and	434+31 (519+8)	585+30 (589+22)	135 (113)	394+29	574+24	567+22	579+19 (826+13)	804+40 (956+20)	139 (115)

Note – see notes to Table I.

Information relating to a 20-year pre-emergency period, (1966-1985), and a 5-year post-emergency period, (1986-1900), has been collected and analyzed. The annual dynamics of indicators of oncologic morbidity and mortality with their follow-up averaging by five-year periods for areas under control (CR^1) and the Brjansk province (BP^2) were studied. It was conducted in three areas.

1. Indicators standardized in age, sex and city village per 100,000 people, by general oncologic morbidity mortality and by separate nosologic forms (Table 1), where distribution in sex, age and city village of the total population of the Brjansk province was used as the standard;

2. Indicators standardized in sex and age per 100,000 people, by general oncologic morbidity mortality separately for urban and rural populations (Table 2);

3. Indicators calculated per 100,000 people, by general oncologic morbidity mortality without being standardized for urban and rural populations, (separately and taken together), in six age groups: 0-14, 15-29, 30-39, 40-49, 50-59, 60 years and over (Table 3).

In addition, geographic prevalence has been considered over 20-year pre-emergency and a 5-year post-emergency periods, all territory under control, distinguishing zones substantially different by levels of oncologic morbidity mortality. Separation of zones was carried out on the basis of calculation of an average indicator over all the years under consideration for each of the Soviets taken separately.

Apart from this, zones with different, averaged over a Soviet, accumulated internal and external doses were separated from 1986 to 1989. A comparative analysis of the dynamics of indicators of oncologic pathology was conducted between 1966 and 1990.

RESULTS AND DISCUSSION

From the data represented in Table 1, it can be seen that standardized indicators of general oncologic mortality in (CR) and (BP) increased somewhat over a 5-year post-emergency period in contrast with the five pre-emergency years. However, analyzing the annual dynamics of mortality indicators, a gradual growth of such indicators is observed over the last

---

[1] CR = Controlled regions.
[2] BP = Brjansk province

ten years without their essential growth over post-emergency years for both (CR) and (BP) (Figure 1a).

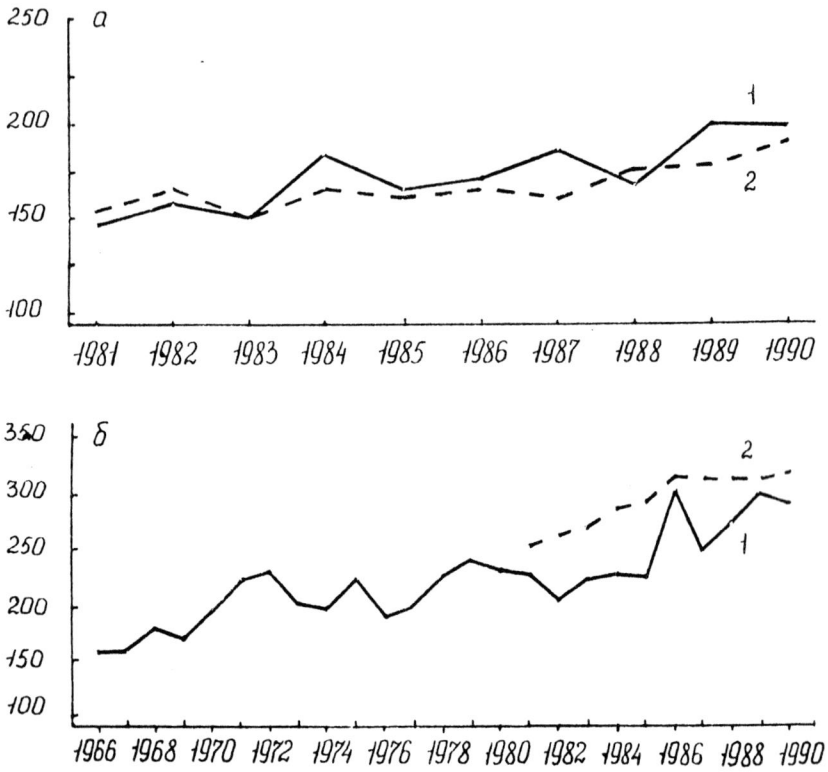

Figure 1. The dynamics of oncologic mortality from 1981 to 1990 (a) and general oncologic morbidity from 1966 to 1990 (b) in areas under control and Brjansk province (2).

Years are along the X - axis; Indicators per 100,000 persons standardized in sex, age and city village are along the Y - axis.

Considering standardized indicators of general oncologic mortality in (CR) and (BP), separately for urban and rural population (Table 2), a more significant post-emergency growth of such indicators in the village is noted. The analysis of the age structure of mortality, (Table 3), showed more than a 20% post-emergency period in the 40-49 age groups, and 60 and over. In cities the increase was noted in the 15-29, 30-39, and 60 and over age groups, whereas (BP) in villages was noted in the 40-49, and 60 and over age groups. It is interesting to note that in 1989 there was a sharp increase in infant and child oncologic mortality in (CR) in the 1-14 year age group, in the city (twice as large as the pre-emergency period) and in the village (1.5 times as large) without a significant increase in the 5-year average level in both (KR) and (BP).

A study of the nosologic structure of oncologic mortality in (CR), (Table 1), has revealed a considerable post-emergency increase in the indicators of a number of cancer localizers such as the oral cavity, esophagus, skin, thyroid, kidneys, prostate, and lymphatic tissue. An increase in cancer of the female genital organs, (mamma, ovaries, uterus, cervix of the uterus), is particularly high. A pronounced maximum level of mortality by the latest two nosologies and oral cavity cancer was noted in 1989 (2 or 3 times higher than the level of the post-emergency period). A the same time, an increase in cancer of the female genital organs, prostate, skin, esophagus is lacking in (BP) following the accident, but is significant in cancer of the oral cavity, kidneys, bladder and in leukemia. The level of mortality in (KR) by such localizers of cancer as the bladder, kidneys, mamma, blood and lymphatic tissue, though exceeding after the accident, never reaches its level in (BP).

The post-emergency level of mortality in cancer of the stomach, liver, pancreas, trachea, bronchi and lungs in (KR) has changed slightly, in regards to cancer of bone tissue and the central nervous system (CNS), some decrease has being observed (Table 1).

A somewhat different picture is observed when analyzing morbidity. In regards to a post-emergency increase in oncologic morbidity, it was 27% in (CR) and 18% in (BP), (Table 1). A rise on the curve of the dynamics, observed in 1986 in (KR). changed for a dip in 1987 resulting in a trend many years after the accident that is practically standing. Values of indicators for (KR) never reach those for (BP) even at the maximum points.

When considering standardized indicators of morbidity in (KR) separately for both urban and rural populations (Table 2), a much higher growth is observed in the city than in the village. The analysis of the age structure of general oncologic morbidity in (KR), (Table 3), has shown that a considerable increase over the post-emergency period was being observed in both the city and village in age groups under 40 and over 60. However, the levels of oncologic morbidity in (BP) after the accident taken for the population as a whole, (city + village), in people over 40 considerably exceeds that in (KR).

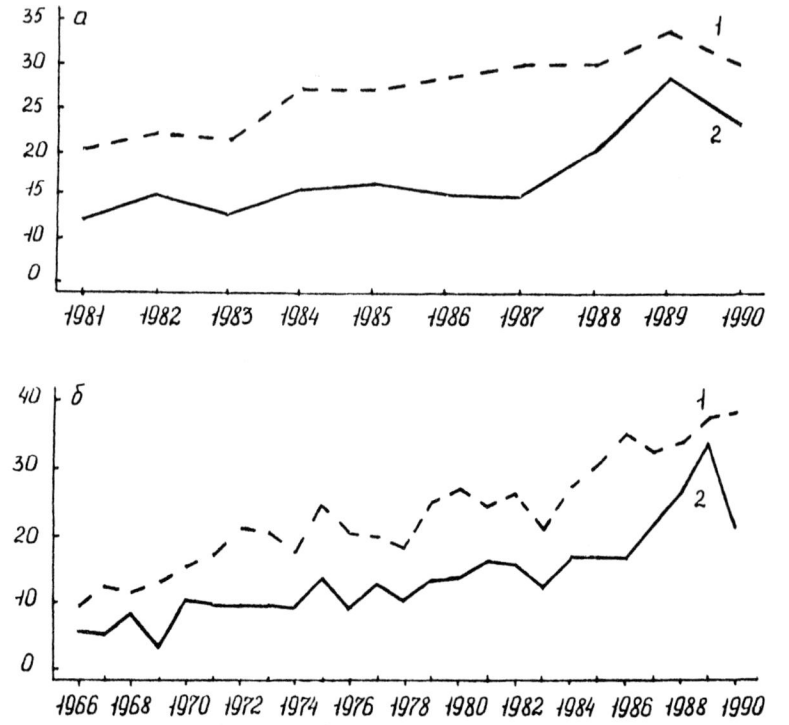

Figure 2. The dynamics of general oncologic mortality from 1981 to 1990 (a) and general oncologic morbidity from 1966 to 1990 (b) in a "high cancer rate" zone (1) and in a "low cancer rate" zone (2) in the areas under control.

Years are along X - axis; indicators per 10,000 people are along the Y - axis.

A study of the nosologic structure of oncologic morbidity, (Table 1), in (KR) revealed the highest increase in indicators from 1986 to 1990 by the following cancer localizers: oral cavity, bladder, thyroid, larynx, prostate, blood and lymphatic tissue. A considerable increase is being observed in cancer of the liver and pancreas, but this has been caused by a sharp decline in the levels of morbidity in these nosologies, which had occurred from 1981 to 1985. In regards to cancer of the female genital organs, an increase in its morbidity after the accident, observed by these localizers, has taken place not only in (CR), but also in (BP), with an increase in cancer of mamma and cervix of the uterus being more significant in (BP). The level of morbidity by cancer of the blood, lymphatic tissue and larynx in (BP) exceeds that in (CR), despite the percentage increase in these nosologies in (CR). The post-emergency level of oncologic morbidity by such localizers as the stomach, kidneys, and cervix of the uterus in (CR) has not really changed, while cancer of the bone tissue and CNS has somewhat decreased.

While analyzing the geographic prevalence of cancer over a 20 year pre-emergency period of all the territory under control, two zones were distinguished: a "high cancer rate zone" and a "low cancer rate zone" (with a population of 33,000 as of 1989 and 26,000 people, respectively). The level of general oncologic morbidity and mortality in these zones differed almost twice . In the pre-emergency, "high cancer rate zone," a trend of the curve of the dynamics of general oncologic morbiity following the accident, (Figures 2a, b) retaines a general pre-emergency character, having a tendency towards gradual growth. A surprising "response" after the accident is seen in the "low cancer rate zone", a sharp increase, (in morbidity by 86%, in mortality by 67%), in comparison with the 5 year pre-emergency period.

Earlier [1], the authors made an analysis of the dynamics of indicators for general oncologic morbidity mortality and some cancer localizers from 1983 to 1985 and 1986 to 1988 in four zones of radioactive contamination (CR), differing by levels of Cs-137 fallout from 4 to 45 Ci km^2. As a result, no relationship was found between the indicators of oncologic pathology and the extent of territory contamination with Cs-137. In addition, post-emergency, "low cancer and high cancer rate zones", did not correspond to either of the zones of radioactive contamination.

A correlation analysis [1] has been carried out to find the relationship between dose levels on the inhabitants of (CR), (a sum of internal and external irradiation from 1986 to 1988, cGy), and indicators of general oncologic morbidity.

Such analysis were conducted for the population of all (CR. Neither of the calculated coefficients revealed any reliable relationship between the values being examined. The numerical values of the correlation coefficients varied between 0.23±0.26 and + 0.42±0.21.

With the goal of elucidating the reason for a sharp post-emergency increase in the level of oncologic pathology in the pre-emergency "low cancer rate zone", the authors divided it, (as well as the "high cancer rate zone"0, into two subzones: a "low dose subzone", (a sum of internal and external irradiation from 1986 to 1989, less than 2 cGy) and a "high dose subzone" (more than 4 cGy). The analysis of the dynamics of general oncologic morbidity and mortality, (Figures 3a, b), in the "high cancer rate zone" showed a lack in essential difference between "high" and "low dose" subzones, while a "low dose subzone" of the "low cancer rate zone", (by 83% as compared with pre-emergency 5-year period), and mortality (by 68%), what is not observed in the "high dose" subzone of the "low cancer rate" zone. That is, a post-emergency increase in the level of oncologic pathology in the pre-emergency "low cancer rate" zone is exactly determined by the "low dose" subzone.

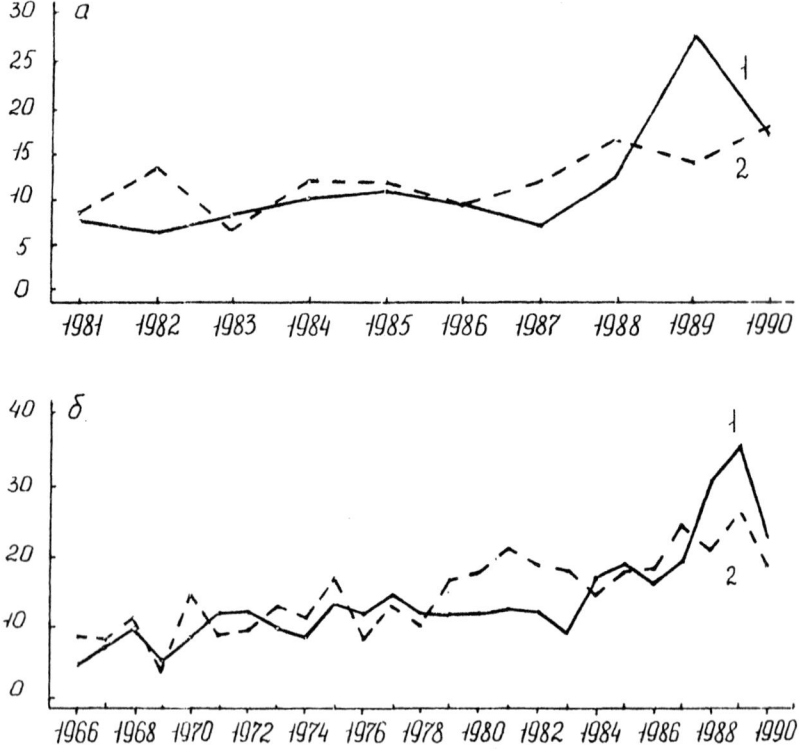

Figure 3. The dynamics of general oncologic mortality from 1981 to 1990 (a) and the general oncologic morbidity from 1966 to 1990 (b) in a "low dose subzone (1) and in a "low dose subzone" of areas under control.

Years are along the X - axis; indicators per 10,000 people are along the Y - axis.

CONCLUSION

The results obtained in this paper give evidence of an increase of oncologic morbidity and mortality after the accident in (CR), 2 or 3 times the level of the 5-year pre-emergency period by separate nosologies and age groups. It should be noted that the increase is determined mainly by 1 to 2 year peaks in the annual dynamics of indicators under examination. A tendency to increase the dynamics of oncologic pathology in the post-emergency period is also characteristic of (BP), where the percentage of the increase in the indicators by some cancer localizers and age groups exceeds that in (CR) after the accident.

The increase in the level of oncologic pathology in areas affected after the accident cannot find an unambiguous explanation now. A period of 5 years is insufficient for radiation to induce carcinogenesis. The obtained results, while finding a correlation between dose levels on inhabitants of (CR) and indicators of oncologic morbidity mortality, as well as while zoning, do not allow an influence of a radiation factor in an oncologic situation in (CR) to be revealed, at least not now.

It seems probable that a complex effect in the population of (CR) of the changed factors in the environment and habitual living conditions, (food-stuffs delivered from outside, disturbance of natural protein-vitamin equilibrium), combined with stress, and increase anxiety of the population, promoted the development of oncologic pathology initiated in the pre-emergency period.

In addition, one cannot help but take into account the mass medical examination of the population of (KP) which was undertaken in the post-emergency period by order of the Ministry of Public Health, resulting in an increase in diagnostication of oncologic diseases, up to 38% in some areas, and an improvement in the quality of pathology diagnosis.

REFERENCES

1. **Troitskaya M.N., Dudarev A.A.** *Radiation Hygiene: Collection of Scientific Works*, St. Petersburg, 1991, pp. 50-54.

CYTOGENETIC EFFECT IN SOMATIC CELLS OF PERSONS AFFECTED BY RADIATION EXPOSURE IN CONNECTION WITH THE CHERNOBYL ACCIDENT

M.A. Pilinskaya, A.M. Shemetun,
A.J. Bondak, and S.S. Dybskij
Ukraine Scientific Center of Radiation Medicine, Kiev

A dose dependent injury of the genetic apparatus of human cells is one of the important aspects of the biological effect of ionizing radiation. This phenomenon has a double meaning. First, it underlies a number of pathological states caused by the mutagenic effect of exposure, second, it is used for biological indication and biological dosimetry of radiation exposure to man by a frequency of chromosome and genetic mutations in somatic and germ cells.

The level of stable and unstable chromosome type aberrations in the lymphocytes of peripheral blood of men is the most widely-spread, well approbated and sufficiently correct biological indicator of an additional radiation effect [1,4,5,9,10].

This cytogenetic criterion is widely used for both personal dosimetry, in the case of any suspected overexposure of men, (in emergency situations, in particular), and for a population indication of additional radiation exposure (e.g. in occupational contacts, in assessment of mutagenic effect of high radiation background, and in the use and testing of nuclear weapons) [3,5,11].

The question that cytogenetic indication of ionizing radiation effect has on man, naturally, acquired a special interest both individually and in the population, in connection with the Chernobyl nuclear power plant accident, (a result of which significant contingents of people were exposed to radiation of different intensity), people who had taken part in the elimination of the effects of the accident and people living in the areas of radioactive fall-outs [2].

The laboratory of cytogenetics Ukrainian Scientific Center of Radiation Medicine (USCRM), founded in January 1987, were only allowed to investigate the exposed contingents 9 months after the accident. Such terms, undoubtedly, restricted possibilities with respect to personal dosimetry due to the elimination of aberrant cells and unstable chromosome aberrations whose frequency, according to the data [6,7] in the first 2 to 3 years after exposure decreases 50 percent per year, on the average.

At the same time, long-lived mitotically inactive lymphocytes might enter into their first post-radiation mitosis in rather late periods after exposure because unstable chromosome injuries could be detected even after many years since acute radiation exposure has been observed in the survivors of the atomic bombings in Japan. The forming of clones of aberrant lymphocytes from stem cells of lymphocyte producing organs injured during exposure is not excluded either.

From this, in a cytogenetic examination of the professional contingents taking part in elimination of the effects of the accident, one would determine only an approximate radiation level by the residual frequency of chromosome type aberrations at the moment of examination.

For the purpose of personal cytogenetic dosimetry 2 groups of exposed people were examined; 55 people with acute radiation syndrome (ARS) in connection with the accidents, patients of the Department of Radiologic Pathology of the Institute of Clinical Radiology of USCRM, and 325 people taking part in the elimination of the effects of the accident and undergoing medical examination and treatment in the clinic of USCRM with different diagnoses.

Blood samples were taken, as a rule, as single samples at different times after cessation of a contact of the individuals under examination with the radiation factor. A bank of chromosome preparations, (about 1000 people) was formed mainly in the period of January 1987 to May 1988 (9 to 24 months after the accident). 25 patients were examined repeatedly with an average interval of 6 to 12 months to study the dynamics of the radiation-induced cytogenetic effect.

Blood was cultivated by a semi-micromethod for 45 to 50 hours which enabled cells whose main mass was in the first postradiation mitosis to be examined. A routine method was used for the painting of the chromosome preparations. A cytogenetic analysis was carried out with encoded preparations. No less than 100 metaphases of every patient meeting the necessary requirement were analyzed.

Chromatid and chromosome type aberrations were taken into account. Chromosome type aberrations such as conjugate fragments, dicentric and ring chromosomes, and symmetric chromosome translocations were considered markers of radiation exposure.

As a result of the cytogenetic examination of the professional contingents the following basic data have been obtained.

In a group of patients with an initial diagnosis of ARS[*] but with a different degree of severity in individuals with the identical diagnosis, an interindividual variability in the level and spectrum of chromosome aberrations was found. Discrepancy between the cytogenetic effect and ARS degree in late periods after radiation exposure could be determined by irregularity of exposure, individual radiosensitivity, different speed of elimination of aberrant cells and, in a number of cases, by hyperdiagnosis. At the same time, probability of detection of cytogenetic markers 12 months after or before the accident increased with a degree of severity of radiation injury.

In 86 patients, with a known dose of exposure, a positive correlation between cytogenetic parameters and the data of physical dosimetry was not always observed. The results of cytogenetic examination spoke, as a rule, of the less intensive exposure which is probably connected with the elimination of chromosome mutations with time.

Among the 239 people with an unknown exposure dose an approximate level of radiation exposure in 220 persons did not exceed 25 cGy, in 35 people it was equal to 25-50 cGy, and in 4 people 60-90 cGy.

The obtained data have confirmed the complexity of reconstruction of individual exposure doses after 12 months or more.

To study the possibility of reconstruction of initial exposure doses by the results of the late cytogenetic examination, a repeated control of people with a high level of radiation exposure is carried out. In the course of a 3 year follow-up of 25 patients an interindividual variability of the dynamics of the radiation-induced effect was found - this included elimination of different intensity, stabilization, and a tendency to a temporary increasing. Studies in this direction are still going on.

The next section of the authors report is a cytogenetic examination of representative population groups living in the areas of radioactive fallouts in the territory of the Ukraine.

---

[*] acute radiation syndrome

It should be noted that implementation of cytogenetic investigations at chronic radiation exposure is complicated by 2 incompatible events: the accumulation of cytogenetic malfunctions in lymphocytes of peripheral blood and stem cells of lymphocyte producing organs; elimination of aberrations with time due to the death of the injured cells and constant renewal of a pool of normal circulating lymphocytes. For computation of a true level of unstable chromosome aberrations induced by prolonged exposure, [13] A. Leonard proposed a so called correction factor in 1986, the quantity of which is determined by the time of exposure within 1 to 40 months and amounts to 1.12 to 9.24 respectively. When using the correction factor, it is assumed that the frequency of aberrant cells decreases exponentially to a 3 year half-life of lymphocytes and that the intensity of the exposure was even all ever exposure.

A number of authors [8,12] in large-scale examinations of professional contingents contacting with ionizing radiation not only showed a reliable increase in the group average frequency of cytogenetic markers of radiation exposure with an increase in its duration but also suggested a coefficient of the yield of dicentrics per 1 cGy at chronic exposure. Such works open the possibilities to carry out not only population cytogenetic indication but also cytogenetic dosimetry at prolonged radiation exposure.

From the point of view of population indication of low intensity radiation exposure the authors examined inhabitants of a number of villages of the Kozelets and Chernigov districts of the Chernigov region, (where short-term concomitant radiation exposure due prevailed due to the fallout of 1-131), and inhabitants of the Norodichi and Ovruch districts of the Zhitomir region (where, in addition to the effect of iodine radionuclides, people were exposed to additional prolonged radiation due mainly to the long-lived Cs-137 radionuclides).

In the Kozelets (May 1988) and twice (in the Norodichi districts) - (from May 1988 to May 1989), groups of children from ages 6 to 16 were examined. Since the end of 1989 children living in Ovruch district have been examined.

Adults aged 16 to 60 were examined in all 4 districts in February 1988.

The group of people examined included people who denied a possibility of deliberate contact with either known, or supposed mutagens, except for ionizing radiation. Blood for cytogenetic analysis was taken under expedition conditions, (during clinical examinations), or when an individual from the areas under control was admitted to the clinic of USCRM. 200 to 500 metaphases were analyzed from each individual.

The results of cytogenic examination of analogous age groups carried out in the agricultural areas of the Ukraine (the Cherkassk region) prior to the Chernobyl accident served as a control to be compared within the data obtained.

In a group of children living in the Kozelets district the average level of aberrant metaphases was 2.62±0.18 percent, (with an amplitude of fluctuations at 0.5±4.5 percent), and reliably exceeded that in the control (1.74±0.19 percent). At the same time, single acentric fragments prevailed among chromosome injuries. Chromosome type aberrants were represented by conjugate fragments only and by a small amount of acentric rings (0.98 and 0.06 per 100 metaphases), whose total frequency differed from analogous control indices (0.47 per 100 cells). Dicentrics and dicentric rings were not recorded (being equal to 0.02 per 100 metaphases in control).

In a group of children living in the Narodichi district a statistically sufficient increase was observed under examination, (unlike the control and Kozelets groups), both in the general frequency of aberrant cells up to 3.78±0.36 percent (with an amplitude of individual fluctuations 1.5±8.5 percent), and in the frequency of chromosome type aberrations (1.88 per 100 metaphases). The latter were represented by conjugate fragments, acentric rings and dicentric chromosomes, which were recorded with a frequency of 1.58, 0.22, and 0.06 per 100 cells, respectively.

Twelve months later, during a repeated examination of a group of children in the Narodichi district admitted to the Department of Endocrinology of the clinic it was ascertained that an average frequency of aberrant metaphases remained at the same level (3.55±0.31 percent), however, the frequency of chromosome type aberrations reliably increased (up to 2.62 per 100 cells) at the expense of acentric rings, dicentric and ring chromosomes amounting to 0.58, 0.19, and 0.19 per 100 metaphases, respectively. A total frequency of dicentrics and rings (0.38 100) considered to be markers of radiation exposure exceeded control level of these mutations by about 20 times (0.02 100) and exceeded their spontaneous frequency by 3 times, which according to review [5] is taken as 0.13 per 100 metaphases.

It should be noted that in the group of repeatedly examined children from the Narodichi district a significant increase was found, (13 times as much in comparison with the Kozelets and control groups), in the frequency of poliploidy metaphases, (up to 0.32 per 1000 interphase and 4.8 per 1000 metaphase cells). This type of injury of the chromosome apparatus of cells, which is rare in populations not burdened by the effect of mutagen factors, is evidence of the malfunction of the process of division, (spindle blockade), in lymphocytes or their predecessors.

In the children of the Ovruch district, (a cytogenetic analysis has not been carried out yet), a tendency is taking shape towards an increase in the frequency of aberrant cells, (up to 3.85±0.32 percent), chromosome type aberrations, (up to 2.38±0.24 percent), as well as in the sum of dicentric and ring chromosomes (0.27 100 cells). The frequency of occurence of poliploidy is higher than in the Kozelets and the controlled areas, but lower than in the Narodichi district (0.07 and 1.54 per 1000 interphase cells).

During an examination of adults of the controlled territories the results of cytogenetic analysis agreed with that mentioned below. Thus, in the inhabitants of Kozelets district a reliable increase in the frequency of aberrant cells over the control level, (2.57±0.28 and 1.45±0.19, respectively), was conditioned by an increase in the frequency of chromosome type aberrations. Though total frequency of dicentrics and rings somewhat exceeded that in the control, (0.12 and 0.05 per 100 metaphases, respectively), it did not differ from the average population index (0.13 100).

A similar picture was observed in a group of examinees from the Chernigov district, where the average frequency of aberrant cells was 2.90±0.29 percent, but chromosome type aberrations were met more often (conjugate fragments - 0.8 100; sum of the dicentrics and rings - 0.15 100).

A specific for the radiation exposure cytogenetic effect was found in the inhabitants of the Narodichi and Ovruch districts where the average group level of aberrant lymphocytes was 2.67±0.29 and 3.56±0.34 percent, respectively, however, the occurence of chromosome type aberrations was reliably increasing - that of dicentrics and rings, in particular (up to 0.44 per 100 metaphases).

One may think that the cytogenetic effect found in population groups of the Narodichi and Ovruch districts, (which is characterized by increase in the frequency of chromosome type aberrations, including dicentric and rings), is induced by radiation exposure.

At the same time, a high level of chromatid type aberrations in all the groups examined (1.73±1.97 100 with 1.05 100 in control and 0.56 100 in average population index) gives evidence of the effect of some mutagen factor of chemical nature identical with all examined agricultural areas of the Ukraine. One may assume that these are genetically active pesticides which are present in the chemical means of plant protection used in the areas under observation. One should not exclude the role of so-called "mutagen stimulation" at the expense of protein deficiency observed in the areas under examination. As it is known, protein deficiency in nutrition does not lead to induction of chromosome aberrations by itself, but it does promote an increase in the sensitivity of the genetic apparatus of human cells to the effect of mutagens.

Thus, the initial results of cytogenetic monitoring of population groups living on the territories of the Ukraine under control after the Chernobyl accident were in positive correlation with the ecological situation there and confirmed a possibility of carrying out population of low intensity prolonged exposure with the help of cytogenetic criteria.

The authors have been trying to assess the cumulative exposure dose of the contingents under examination using Lloyd's formula [12] for the yield of dicentrics at chronic occupational radiation exposure: $y=(2.2\pm0.94).10^{-4}D$, (ignoring the quadratic portion of the equation). According to the computations made, an approximate exposure does for the Narodichi district may reach 18.9 rem, (for 3 years), for the Ovruch district - 20.2 rem what slightly exceeds the results of physical dosimetry received from the Central dosimetric register of the Department of Dosimetry and Radiation Hygiene headed by professor I.A. Lichtarev (8.6 and 8.9 rem, respectively).

New methods for determining somatic mutations in man, glycophorine test and fluorescent hybridization of metaphase chromosome in situ with DNA probes, enables the detection of dependent stable injuries of the genetic apparatus of cells which do not disappear with time, and seem to be very promising for genetic indication and dosimetry of radiation exposure in late periods after the accident. A scientific cooperation with Livermore Research laboratory (USA) has begun to use these methods with respect to the Chernobyl contingents.

REFERENCES

1. **Bochkov N.P.** *Chromosomes of man and radiation.* - M. 1971.
2. **Iljin L.A., Balonov M.I., Buldakov L.A.** *Medical Radiology.* 1989 - pp. 59-81.
3. **Kudritskij J.K., Karpov V.I.** Review information of All-Union Institute of Scientific and Technical Information (UISTI), *Hygiene series.* - 1984 - Issue 3 - pp. 31-39.
4. **Leonard A.** *Cytology and genetics.* - 1986 - Vol. 20, No. 2, - pp. 115-121.
5. **Pjatkin E.K., Baranov A.E.** Summary of Science and Technology. *Radiation Biology.* - Vol. 3 - pp. 103-179.
6. **Bender M.A., Awa A.A., Brooks A.L.** et al. *Mutat. Res.* - 1988 - Vol. 196 - pp. 103-159.

7. **Evans H.J., Woodhead, A.D., Shellaberger C.J.** et al. *Assessment of Risk from Low lever Exposure to Radiation and Chemicals.* - New York, 1985 - pp. 429-451
8. **Evans H.J.**, Banbury Report 13, Ed. B.A. Bridges et al. *Indicators of Genotoxic exposure.* Spring Harbor, pp. 325-336.
9. **Evans H.J., Buckton K.E., Hamilton G.A., Carothers A.** *Nature.* - 1979 - Vol. 172 - pp. 531-534.
10. **Littlefield L.G., Joner E.E., DuFrain R.J.** et al. *The Medical Basis for Radiation Accident Preparedness.* - New York, 1980 - pp. 375-390.
11. **Lloyd D.C.** *Biological Dosimetry: Cytometric Approaches to Mammalian Systems* - New York, 1984 - pp. 3-14.
12. **Lloyd D.C., Edwards A.A., Prosser J.S.** *Doses in Radiation Accident Investigated by Chromosome Aberrations Analysis* (VX. A Review of Cases Investigated. 1985, National Radioprotection Board Report NRPB). - Chilton, 1986.
13. **Hoyd D.C., Pussot R.J., Reeder E.J.** *Mutat. Res.* - 1980 - Vol. 72 - p. 523-532.

HYGIENIC EVALUATION OF THYROIDAL RADIATION DOSES IN INHABITANTS OF THE UKRAINE FOLLOWING THE CHERNOBYL ATOMIC POWER STATION ACCIDENT

A.Ye. Romanenko, I.A. Likhtarev,
N.K. Shandala et al.
Ukraine Scientific Center of Radiological Medicine, Kiev

The functioning of any nuclear reactor is accompanied by injecting the fragments of iodine radioisotopes into the environment [2]. So, the cumulative radioactive iodine effluents in the atmosphere in 1980 from all nuclear power stations were estimated at 20 TBq. In emergency situations, the iodine effluent in the atmosphere becomes one of the main factors of human radiation [2,5]. During the accident at the radiochemical plant in Windscale (England) in 1957, about 700 TBq of ^{131}I were released into the atmosphere, and during the course of the accident at Three Mile Island nuclear power station (USA) in 1979, the pollution of the air with the radioactive iodine was estimated to be equal to 0.6 TBq [3]. During the Chernobyl accident, the nuclear power plant discharge of ^{131}I from the data of May 6, 1986 came to 270 PBq [3].

In the first stage after the Chernobyl accident (up to July 1986), ^{131}I was one of the leading radiation sources among the population. To estimate the radiation condition stemming from radioactive iodine discharges in Ukraine work on creating a data bank on thyroid gland (TG) radiation was carried out. The results of this work have formed the basis for the evaluation of individual absorbed doses of TG radiation due to ^{131}I in children and adults in the Ukraine.

MATERIALS AND METHODS

The findings of TG radioactivity determination in more than 150 thousand Ukrainian inhabitants from May to June of 1986 are at the root of the authors studies. They were performed by special dosimetry teams, (approximately 100), with the participation and methodic guidance of specialists of the Leningrad Research Institute of Marine Hygiene. (L.R. Romanov, G.L. Moroz, A.N. Kovtun et al.). At the preparation stage, the systematization and formalization of available information was performed and for this purpose the dosimetry iodine bank was created. To run it, personal computers and data base control system "DBASE 3+", "FOXBASE +" and "DBASE 4%" were used. In addition a program envelope was developed which allows one to realize the necessary set of input-output operations and statistical analysis of obtained results.

The block of passport data, results of instrumental determinations, design data on the iodine content in the thyroid gland (TG), and doses of TG radiation, information on migration of an individual during iodine period of the accident, reference library of source information and the block of ancillary data became the principal structural units of the dosimetry iodine bank. The information unit of the iodine data bank is a formalized individual chart of dosimetry monitoring of TG radiation comprising an individual code of a person, and administrative code of locality at the instant of the accident, passport data of a person, data on the site of source information storage, information concerning dates and evacuation points, radiation dose calculation results from various models of iodine intake and others.

The initial calculation of TG radiation doses was performed in compliance with conventional procedures [1,2]. At the first stage of the calculation of radiation doses, the most conservation model of 131I "single uptake" into the human organism was used. It gives somewhat higher values of TG radiation doses as compared to the more realistic dosimetry model used in subsequent calculations that take individual dynamics of ^{131}I intake into account. The analysis demonstrated that the mean correction factor, when passing to a more realistic model, is within 1.2 - 2 times. However when analyzing the results in this paper, the authors consider it to be more expedient to be guided by dose evaluations performed from a conservative model since, at present, the contribution of iodine short-lived isotopes into the individual dose of TG radiation is not conclusively established. The estimation of their contribution can demonstrate that the real doses, to which some contingents of evacuated people and urban inhabitants of the Ukraine were exposed and which were calculated from the model of long-term intake of ^{131}I intake, are higher. In addition, at the creation stage of the dosimetry iodine data bank, the work on forming the contingent of people subjected to prolonged specialized endocrinologic monitoring, primarily children, was carried out. In order to encompass the wide range of the population exposed to radiation with specialized observation, practicing physicians were informed of children's TG radiation doses calculated from the conservative model which is extremely important from a medical point of view.

RESULTS AND DISCUSSION

Figure 1 shows average TG radiation doses in children and adults from various regions of the Ukraine exposed to ^{131}I emissions. The results are ranged as the radiation doses diminished, (these figures being of children not older than 7 years of age [born from 1979 to 1986]), at the moment of the accident, they are a critical group of residents both from the viewpoint of radiation doses and TG radiation sensitivity. The highest absorbed dose, (from 300 to 700 cGy), were taken from children in the Narodichsky and Ovruchsky regions of the Zhitomir area (oblast') and in the Pripyat' and Polessky regions of Kiev area. The changes in TG average radiation doses according to the places of residence in other age groups corresponds to similar values typical for children at the age of 7 in general. At the same time, one can clearly see the regular, well-known, decrease in the dose loads with the increase in age [2]. Thus, the average TG doses of children from age 7 to 15, (born during 1971 to 1978), in the above regions at the time of the accident are about 2.5 times lower as compared with those in the 0 to 7 age group.

However, when analyzing dose loads, one should take into account that distribution of TG radiation doses is of lognormal type. Therefore, the extent of iodine exposure should be judged from the distribution of children and adults in dose groups, (see the Table).

The table lists the data for 19 regions and settlements of the Kiev, Zhitomir and Chernigov areas which suffered the most from iodine exposure. These data are based on a direct measurement of iodine content in TG in 20 to 90% of the children and 1 to 10% of the adults in the above regions. For that reason, the regularities of this distribution can be spread with high degree of reliability over all the residents of the aforementioned territories. When estimating the size of dose groups, age standardization of population was performed.

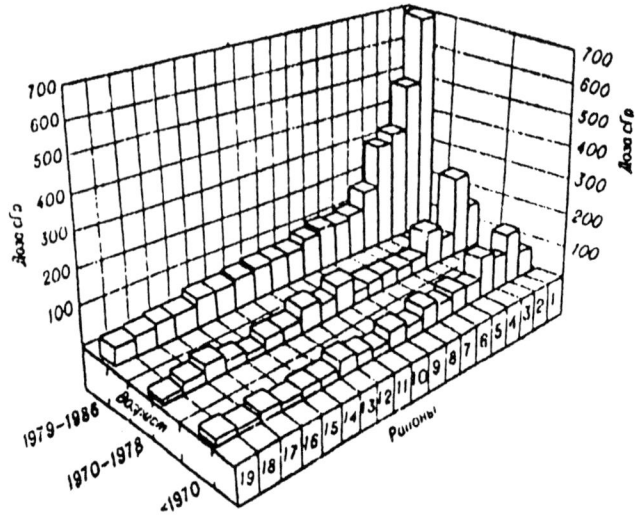

Figure 1. Average TG radiation doses of inhabitants of the Ukraine according to age and place of residence. 1.-Ovritchsky region. 2.-Narodichsky region. 3.-Town of Pripyat'. 4.-Polessky region. 5.-Chernobyl region. 6.-Ivankovsky region. 7.-Makarovsky region. 8.-Vyshgorodsky region. 9.-Borodyansky region. 10.-Chernigovsky region. 11.-Kozeletsky region. 12.-Repkinsky region. 13.-Town of Ovrutch. 14.-Vinnitsa Area (Oblast'). 15.-other regions of Zhitomir Area. 16.-Olevsky region. 17.-Korostensky region. 18.-Town of Chernigov. 19.-Kievo-Svyatoshinsky region.

Judging from the size of the group of children with TG radiation doses with more than 200 cGy (children with increased radiation risk), the group that suffered the most from iodine are the inhabitants of the Narodich and Ovruch regions of the Zhitomir area, Polessky, Ivankovsky, Chernobyl regions and the town of Pripyat' of the Kiev area, as well as the Chernigov region of the Chernigov area. Yet, at the same time, one should take into account that for these regions, with the exception of the Chernobyl region and Pripyat', the prolonged ^{131}I intake is typical and, hence, considering the dynamics of its content in air and food-stuffs, the values of radiation loads on TG can be slightly lower.

Age-Group Distribution of Inhabitants of the Regions in the Ukraine Most Severely Suffered from Radioactive Iodine Exposure

Regions	Size of dose-age group, %							
	Born in 1970–1986				Born prior to 1970			
	Dose, G_u							
	0–0.3	0.3–1	1–2	2	0–0.3	0.3–0.1	1–2	2
Narodichski	11.0	25.1	20.1	43.8	20.8	37.3	22.7	19.2
Ovrutchski	32.6	41.1	15.5	10.8	38.9	47.0	10.8	3.3
Town of Ovrutch	34.0	52.9	10.0	3.1	64.3	32.5	2.4	0.8
Korostenski	48.9	40.6	7.6	2.9	77.2	19.1	2.7	1.0
Olevski	56.2	33.7	7.0	3.1	85.3	13.9	0.8	0
Zhitomir area (oblast')*81.3		11.1	1.0	6.6	69.2	26.9	3.0	0.9
Chernigovski	43.7	33.1	12.5	10.7	64.9	26.3	5.8	3.0
Repkinski	46.8	34.0	11.1	8.0	76.3	20.4	2.6	0.7
Town of Chernigov	60.6	32.7	4.9	1.8	91.0	7.0	1.1	0.9
Kozeletski	42.2	39.0	12.2	6.6	84.4	14.6	0.8	0.2
Chernobylski	54.5	25.0	9.9	10.6	58.3	36.3	3.9	1.5
Town of Pripyat'	44.6	29.4	10.9	15.1	40.5	40.2	10.1	9.2
Polesski	37.2	27.9	14.7	20.2	37.0	37.8	14.9	10.3
Ivankovski	56.8	24.8	8.3	10.1	58.1	32.5	4.3	5.1
Borodyanski	51.9	34.7	7.7	5.7	80.0	16.8	3.2	0
Makarovski	35.9	46.8	8.8	8.5	64.0	29.4	6.4	0.2
Vyshgorodski	37.3	43.6	11.8	7.3	50.9	33.0	11.5	4.6
Kievo-Sviatoshinski	79.6	15.6	2.4	2.4	83.6	8.5	7.9	0
Vinnitsa area	51.1	35.3	8.1	5.5	87.5	9.6	1.9	1.0

* Predominantly based upon data concerning Zhitomir, Luginsk and Novgorod-Volynski regions of Zhitomir area

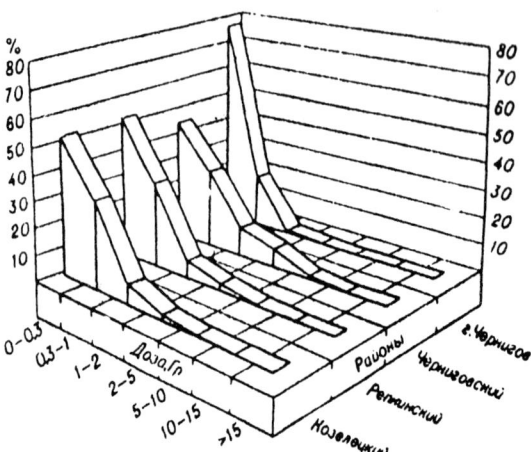

Figure 2. Distribution of TG radiation doses in inhabitants of the Chernigov Area born during 1970 to 1978.

Figures 2, 3, 4 give a more detailed distribution of TG radiation doses in the most "measurement-embraced" group of residents (children born from 197 to 1978).

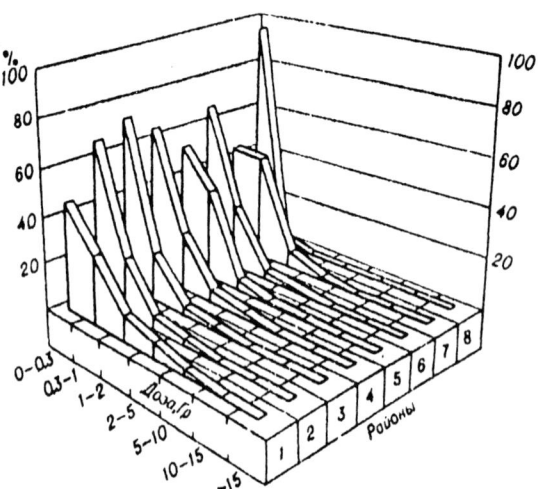

Figure 3. Distribution of TG radiation doses in inhabitants of the Kiev area born during 1970 to 1978.

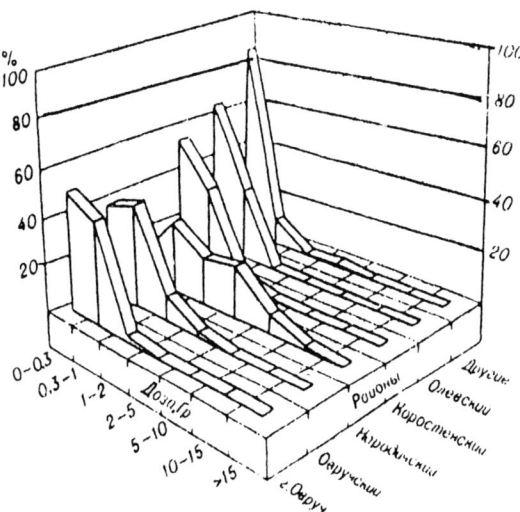

Figure 4. Distribution of TG radiation doses in inhabitants of the Zhitomir Area born during 1970 to 1978.

Now, in evaluating possible remote sequels due to TG radiation, one should note that forecasts were made for both children, (they are considered to be a critical contingent from the viewpoint of expected radiation-induced TG pathology with consideration for high radiation sensitivity of TG in children 7), and also for the expectancy of adults.

The estimation of cumulative TG radiation was made in regions where there is the most representative group of residents with established individual absorbed doses. Therefore, from the analysis of areas given in the table, the Vinnitsa area and some regions in the Zhitomir area were excluded, as a rule, because relatively low individual doses are typical in these regions. The cumulative thyrooncogenetic doses of TG radiation of children and adults were 23,000 and 357,000 man.Sv, respectively. They were calculated for residents of the above areas the population of which, according to data of January 1, 1986, was 1,277,000.

The risk factor of induction for all types of TG malignant tumors was 46 tumors per 10,000 man.Sv for children and 8 for adults (the level of incurable malignant tumors was taken to be 10%).

From this, for the population in question, we forecast a possible emergence of radiation-induced TG tumors in children at the level of 1060 cases, (106 incurable ones), and in adults-285 cases (29 incurable ones).

The given absolute estimation of possible number of thyrooncogenetic effects should be compared with a spontaneous level of TG carcinoma morbidity. However, there is little data regarding the areas in question. The use of world observation data can be made with very seious provisos since the variability of spontaneous thyrooncogenesis in different countries is extremely high. In the literature, [6], the spontaneous level of clinically detected cases of thyroid carcinoma was estimated at 2.6 cases per 100,000 individuals a year in males, and 8.2 cases in females. The authors can take, as a certain "plausible" estimation of TG spontaneous tumors the value of 2 cases of thyroid carcinoma per 100,000 individuals a year. In this respect, from the data [7] and taking into account that the minimum latent period of radiation-induced tumors is 5 years after radiation exposure [8], the spontaneous level of morbidity due to TG carcinoma over the remaining life-span for children (born from 1970 to 1986) may account for the 300 malignant tumors. Sex differences were not taken into account and the period of "life under risk" was assumed to be 60 years [7].

Considering life expectancy in the nearest 60 years, it is possible to have a three-fold rise in the spontaneous morbidity level due to thyroid carcinoma in children in areas under review.

The studies performed makes it possible to draw the following conclusions:

1. The critical group of the population who suffered from ^{131}I radioactive emissions due to the Chernobyl accident are children. The highest levels of TG radiation are typical for children of under-school age (born from 1979 to 1986). The average absorbed doses in school-year children (born from 1970 to 1978), are 1.5 to 3 times lower than in pre-school children; in adults this figure is 2.5 to 3.5 times lower than in children born from 1970 to 1986.
2. The distribution of TG radiation doses is of lognormal character, therefore, when estimating the rate of ^{131}I exposure to residents who suffered due to accident, the data on dose-age distribution should be used but not the average values of absorbed doses. This approach allows one to optimize the monitoring of the TG condition with respect to the regional features of its radiation.
3. Estimation of possible remote radiation thyrooncogenetic pathology made on the basis of TG radiation findings calculated with the help of the most conservative model of single-dose

131I intake into an organism has demonstrated that, for the remaining expectancy of the areas in question, 1060 TG carcinoma cases in children and 300 cases in adults are expected.
4. To refine the hygienic evaluation of thyroidal radiation exposure in the inhabitants, it is necessary to correct the radiation doses with consideration for individual dynamics of ^{131}I intake into human organism as well as the contribution of short-lived isotopes into cumulative radiation doses.

REFERENCES

1. **Z.S. Arefyeva, V.I. Badyin, Yu. I. Gavrilin** et al. *"Manual for Estimation of Thyroidal Radiation Exposure with Radioactive Isotope Intake into Human Organism"*. Ed. L.A. Ilyin, Moscow, 1988 (in Russian).
2. **L.A. Ilyin, G.V. Arkhangelskaya, Yu. V. Konstantinov** et al. *"Radioactive Iodine in the Problem of Radiation Safety"*. Moscow, 1972.
3. Information on Chernobyl Nuclear Power Station Accident and Its Sequelae Prepared for IAEA. *Atomic Power*, 1986. Publ. 61, N5, pp. 301-320.
4. Ionizing Radiation: Sources and biological Effects: *UNO Scientific Committee on Nuclear Radiation Effects.* Report to UN General Assembly 1982, N.Y., 1982, v.I.
5. **Yu. I Moskalev.** *"Metabolism Kinetics, Biological Effects of Radioactive Iodine Isotopes"*. Ed. Yu. I. Moskalev, Moscow, pp. 7-28.
6. **Ezaki H, Ishimaru T., Hayashi Y., Takeichi N., Gann. Monogr.** *Cancer Res.* 1986, v. 32, pp. 129-142.
7. Induction of Thyroid Cancer by Ionizing Radiation: Recommendation of the National Council on Radiation Protection and Measurement: *(NCPR Report No 80)*. - Bethesda, 1985.
8. **Zeighami A.E., Morris M.D.** *Hlth Phys.* - 1986 - Vol. 50, N1, pp. 19-32.

Health Status of the Adult Population in the Western Districts of the Bryansk Area in 1989

R.N. Turaev
Scientific Research Institute of Roentgenology
and Radiology, Moscow

The health status of the population in the Bryansk area is studied within the framework of the annual clinical examination of people who were subject to radiation.

In 1989, 148,824 people lived in five regions of the western territories of the Bryansk area, (which made up 10% of the area's population), in the regions of "strict" control - 105,744 people (table 1).

The demographic processes were studies in the area. According to statistic materials the reduction in the general death-rate of the population and of the infant death-rate has been noted since 1985. Some decrease in the birth-rate was registered, however, it is impossible to come to any conclusions now as to the reasons for these phenomena. The study of the dynamics of the demographic processes in the western regions at least during the five coming years would allow the demographic indices of certain regions to be estimated and make preliminary conclusions about the reasons for the demographic situation's change. It must also be noted that the demographic processes in the regions depend on the outflow of young people to a great extent and also mothers with children, which cannot but tell on the general demographic situation. The general data about the demographic situation are given in table 2.

Table 1. Population numbers in the controlled regions on January I, 1990.

Region	Number of inhabitants	No. of inh-ts on controlled ter-s				participants of liquidation
		adults	children	pregnant	total	
I. Klintsovsky	29,146	3,544	597	38	4,141	
2. Krasnogorsky	22,700	10,260	3,480	127	13,470	4
3. Novozybkovsky	62,396	47,489	13,107	684	60,596	28
4. Gordeevsky	16,503	10,208	2,926	301	13,123	4
5. Zlynkovsky	18,079	10,864	2,967	121	14,375	2
Total	148,824	82,356	23,076	1,156	105,705	38

Region No.	Evacuated	Under observation	
		total	all-Union level
I.		4,141	635
2.		13,474	3,611
3.		60,624	13,819
4.		13,127	3,230
5.	I	14,378	3,091
Total	I	105,744	24,271

Table 2. Dynamics of demographic processes in the Bryansk area

Index	1985	1986	1987	1988	1989
Number of population (thnd)	1475.7	1473.3	1474.4	1476.0	1470.8
Birth-rate per 1000 people	15.3	16.3	15.9	15.1	14.2
General death-rate per 1000 people	13.2	11.3	11.8	12.4	11.9
Infantile death-rate per 1000 newborn children	18.4	17.2	18.6	17.4	14.1
Natural increase per 1000 people	2.1	5.0	4.1	2.7	2.3
Average length of life, years			69.8	68.4	70.8

The Leningrad Scientific Research Institute of Traditional Hygiene has, at the present time, made calculations of average doses of radiation of the thyroid gland with radioactive iodine, as a result of the accident; depending on the zone of observation and patient's age they average from 10 to 250 bar. The content of radioactive caesium in the organism has also been counted. On the basis of these calculations, the radiation doses of the population were estimated and recommendations were given concerning the settling-out of the citizens of the places where the dose will be exceeding 35 bar till 2056.

Despite the dosimetric measurements, their high trustworthiness and the matching of the results of measurements made by different organizations, the distrustful attitude of the doctors working in the regions and that of the population towards the information about the radiation situation still remains, which is connected with the insufficiently high level of radiological knowledge of the doctors and, in the author's opinion, with the artificial aggravation of the stress situation by a number of local experts and mass media.

ORGANIZATION OF CLINICAL EXAMINATIONS AND THE NETWORK OF CLINICS

The clinical examination of the population is conducted by active, (visiting enterprises and offices), and passive, (patient's visiting clinics), methods. it is conducted in the clinics of the central regional hospitals (CRH) which were mostly built during the last 20 years and which allow to make the necessary diagnostic examination of patients and provide them with a treatment.

The percentage of the population covered by the clinical examination is rather high. All pregnant women are examined, practically all (91-98%) children too. The examination of the adult population is worse, especially in the regions where many of the elderly village citizens live. The decrease in the percentage of the population under examination in 1989 in comparison with 1988 should also be noted. The most considerable decrease in number of those under the examination took place in the Gordeevsky and Klintosovsky regions (table 3).

Table 3. Clinical examination conducted in 1988 and 1989

Region	Supposed to be examined		Examined		% of the population	
	1988	1989	1988	1989	1988	1989
Klintosovsky						
total	4,163	4,141	3,981	3,397	95.6	82.0
adults	3,522	3,544	3,376	2,806	95.8	72.9
children	641	597	618	589	96.5	98.0
pregnant	63	38	63	38	100	100
Zlynkovsky						
total		14,375		11,993		83.0
adults		10,572		9,129		86.3
children		2,910		2,864		98.4
pregnant		134		134		100
Gordeevsky						
total	12,937	13,123	11,857		91.6	90.6
adults	9,916	11,210	8,836	6,923	89.1	61.8
children	2,791	2,915	2,791	2,680	100	91.9
pregnant	230	301	230	301	100	100
Krasnogorsky						
total	13,895	13,470	13,637	11,419	98.1	83.1
adults	10,392	10,260	10,327	7,962	99.3	77.6
children	3,310	3,480	3,310	3,457	100	95.1
pregnant	193	127	193	127	100	100

Table 3 (Continued)

Novozybkovsky						
total	48,910	60,596	44,366	51,063	90.7	84.3
adults	37,665	47,480	33,136	38,409	88.0	80.9
children	11,245	13,107	11,230	12,654	99.9	96.5
pregnant	714	750	714	750	100	100
Western region						
total	79,905	50,705	73,841	51,063	97.7	74.3
adults	61,495	82,356	55,675	65,229	90.5	79.2
children	17,987	23,076	17,877	22,244	99.4	96.4
pregnant	1,200	1,350	1,200	1,350	100	100

During the examination, only the fact of the revealed illness is stated. Not all the patients with revealed illnesses are put under clinical examination and not all are given a preventive treatment.

MEDICAL PERSONNEL

Providing preventive medical and treatment clinics with medical personnel despite measures taken by the Ministry of Health of the RSFSR and the Health department of the Bryansk executive committee is lower on the whole than in the area and in the RSFSR.,

In comparison with 1985, the number of doctors in the Bryansk area increased by 831, (in 1985 to 4512, in 1989 - 5343), and the number of junior medical workers, by 2793, (in 1985 - 12,201, in 1989 - 14,994). The number of medical workers in the western regions of the Bryansk area has also increased. Presently 428 doctors and 1337 junior medical workers work in these regions, which makes 8% of doctors and 9% of junior medical workers from the total number of these specialists in the area (10% of the area's population live in these regions).

For 4 years 322 doctors arrived in the western regions i.e. 29% of the total number of young experts sent to the area; since 1986 211 doctors have left; thus, the increase in doctors in these regions was 11 people. The situation with junior medical workers is somewhat better than that with doctors.

One of the reasons for the outflow of the doctors and their refusal from the assignment is the difficult housing situation; on January 1, 1990 180 medical workers were on the waiting list for a flat in the western regions.

The staff list of the western regions must have 1551.5 junior medical workers; on January 1, 1990 there were 1428 posts (1337 people); there was a shortage of 123 workers.

In connection with the admission of school graduates to medical colleges in the controlled regions without entrance examination recently providing with junior medical workers has improved.

As a results of the small influx of doctors in Zlynkovsky, Gordeevsky, Krasnogorsky and Novozybkovsky regional clinics are not staffed with specialists, (therapeutists, endocrinologist, ophthalmologist). The examinations are not always made thoroughly due to this lack of local specialists. Complete blood tests are made only for children (and not always).

In connection with the shortage of qualified specialists, groups of specialists from scientific research institutes and central preventive and therapeutical medical clinics of Moscow are sent to CRH. Patients go with the recommendation of local health services or go independently to the medical organizations of Moscow and Bryansk.

Table 4. Use of special methods in CRH of the western regions in 1989

Region	X-ray research	fluorography	ultrasound research	endoscopy	HRM
Klintsovsky	16,209	5,457	1,291	2,505	1,328
Krasnogorsky	2,914	3,235	2,697	500	1,450
Novozybkovsky	6,894		11,449	1,121	
Gordeevsky	7,611		891	201	
Zlynkovsky	4,767		1,048	51	
Total	38,364	8,692	17,375	4,386	2,778

Recently, the use of special research methods has been extended and as a result, the diagnostic quality has improved and the radiation load on patients has been reduced. Since the middle of 1989, fluorography has ceased to be used in the western regions. In 1989, ultrasound scanners were used in all the regions to examine children's thyroid glands. The endoscopic research method is not widely used in all the regions, yet. Klintosovsky and Krasnogorsky CRHs have HRM (human radiation meters), however, the data from the research made are not written in dosimetric cards, which does not allow one to find out the individual doses of the population. Although in Novozybkov, in the laboratory of the Leningrad Scientific Research Institute of Radiation Hygiene, there is a HRM. CRH dose not send patients for examination and does not register the data either. Information about the use of special research methods is given in table 4.

DISEASE RATE

In connection with the fact that a number of the population in certain regions under observation is small and that the regions often changed their administrative structure since 1986, the calculations of the standardized indices of some forms of diseases are made for the whole western region.

Table 5 gives information about some forms of disease of the population of the western regions and of the Bryansk area on the whole.

Analysis of the statistical information about the disease rate in the area enables preliminary conclusions to be made that, on the whole, the disease rate in the area remained on the previous level from 1988 to 1989, the disease rate in the western regions has increased, the disease rate in the controlled regions has also remained on the previous level, however, the absolute figures of the disease rate in the controlled region correspond with the disease rate in the western regions in 1989.

The analysis of the rate of certain disease shows that the infectious disease rate in the western regions has decreased during a year by 65%, while, on the average, in the area and in the controlled region it has increased. This indicates the strengthening of the work of SES bodies and the control over the sanitary and epidemiological situation.

The general rate of neoplasm disease in the area remains stable, a small increase is noted in the western regions and an increase by almost twice in the controlled region, which is connected with the participation of highly-qualified specialists in the examination who have revealed patients who have long been suffering from benign and malignant neoplasms.

Table 5. Indices of disease rate in the population of the Brjansk area (per 100,000 people) in 1988 and 1989

Disease	Area 1988	Area 1989	Western regions 1988	Western regions 1989	Controlled region 1988	Controlled region 1989
All kinds of dis-s	56327.0	56367.7	45745.9	63212.1	61535.7	62009.5
Infectious diseases	1631.2	1875.3	2171.4	1413.9	620.1	1116.7
Neoplasms	759.7	746.7	748.5	845.0	714.3	1268.1
Dis-s of endocrine system, metabolic & immunity diseases	330.8	625.5	604.7	1036.8	142.9	8716.1
Theriotoxicosis	24.8	33.0	42.2	37.2	13.0	15.8
Blood diseases & dis-s of blood-making organs	10.4	25.6	12.1	121.2	26.1	12.6
Mental disorders	924.6	1155.6	1458.2	1885.7	311.7	334.4
Dis-s of nervous system and of organs of sense	5919.1	6657.3	2665.8	9651.1	10698.0	11627.7
Glaucoma	28.4	33.0	42.2	37.2	13.0	15.8
Dis-s of bloodstream system	1685.1	1901.5	1971.6	3221.6	1373.6	1555.2
Dis-s of respiratory organs	24052.0	20998.7	14575.3	17535.9	31113.6	21012.6
Dis-s of digestive system	2952.8	1724.7	1926.8	2146.3	1409.0	1362.8
Dis-s of urino-genital system	2431.1	2414.9	1778.6	3599.1	4529.2	3119.9
Skin diseases and dis-s of hypodermic tissue	4536.7	4346.7	3490.1	4027.1	6678.6	2561.5
Dis-s of osseous-muscular and conjunctive tissues	3589.8	3376.5	3143.0	4066.7	2230.0	1353.3

The strengthening of the endocrinological service led to the increase of two times the rate of disease revealed in both the area and in the western regions. In the controlled region, the rate of endocrine disease revealed has increased by practically 60 times which was caused by the special endocrinological examinations conducted with this purpose for the first time. It is this increase that told on the average area index. It must be noted that the rate of revealed theriotoxicosis in the western regions has decreased as the main bulk of such patients were revealed during the previous years.

It is only since 1989 that high-level blood tests were made in the western regions which led to an increase in the rate of diseases the blood and blood-making organs. A more detailed analysis of the results of the experts' research showed that mostly iron-lacking anaemias were diagnosed, which is connected with the aggravation of the food situation.

The analysis of the rest of the indices also shows that the use of modern methods of diagnostics and the involvement of highly-qualified specialists for examinations sharply increase the rate of diseases. Therefore, the dynamic observation of the population for several years would be of a certain scientific and practical interest.

Table 6. Dynamics of the rate of thyroid gland diseases revealed among the adult population of the western and controlled regions of the Bryansk area per 10,000 people in 1986-1989

Form of disease	1986	1987	1988	1989	1989 controlled region
Simple goiter	109	116	117	115	179
Nodal goiter	102	124	108	120	135
Initial hypotheriosis	0.3	1.2	1.4	1.0	-
Chronic therioiditis	0.4	2.7	4.0	9.3	10.4
Hyperplasia	1,628	1,768	1,415	1,974	1,453
Population examined	61,227	56,492	67,889	67,430	14,307

The controlled territories were subject to the fall-out of radioactive iodine and the population received certain doses of radiation of thyroid gland. Because of possible mistakes in establishing the amount of doses, the insufficient knowledge about the collaboration of a region's endemicity and the organ's radiation, the research of the state of thyroid gland requires a lot of attention. During the pre-accident period, little attention was given to the study of the state of this organ. Presently, when experts' examinations are made, a great number of deviations in the state of the thyroid gland is revealed. However, when the state of thyroid gland disease was studies in the Krachevsky region, practically no deviations of the population of the western regions were revealed in comparison with the controlled region (table 6).

Table 7. Spreading of malignant tumors of the thyroid gland in the western and controlled regions

Region	1985	1986	1987	1988	1989
Western regions	2	8	24	15	6
Controlled region	-	2	1	1	9
Area	22	49	66	68	78
Area - western regions	20	39	41	52	63

In all the regions a large number of theriopathies and malignant tumors of the organ was revealed, which required more surgical operations (table 7).

The information about the dynamics of diseases shows that there was no increase in number of nodal and diffusive forms of goiter. The increase in therioiditises was a result of the use of ultrasound research.

The analysis of endocrinologists' activity shows that the endocrinological service does not work actively enough. Thus, in 1989, 54 people were hospitalized to be operated on to the area endocrinological clinic; at the same time there is a waiting list for hospitalization to the scientific radiation research institute. There is an obvious shortage of endocrinologists in CRHs. Only 7.75 posts are occupied instead of the fixed 12.5 in the western regions.

An endocrinological department was founded in Novozybkovskaya CRH but it does not function properly because it is not manned enough and because of a poor laboratory service. These shortcomings cause a considerable flow of patients to medical organizations of Moscow.

The oncologic disease rate in the Bryansk area was higher than that in the republic even before 1986. It should be noted that in 1988 it was lower in the controlled regions than in the area and in the controlled region.

The increase of disease in the population of the western regions registered even before 1986 and 1987 is connected with the overall clinical examination of the population. Later, the decrease of the oncologic disease rate was noted among the population of all the regions except Klintsovsky, where the smallest number of those under observation live at a ratio to the total number of the citizens of the region, and at the same time, the contingent of people examined is increasing in addition to the compulsory one. The rate of oncologic disease actively revealed in these regions reached 36% while the average areal index is 20.6%, which characterizes the improvement of the quality of oncological diagnostics in the western regions.

In 1987, the increase of the rate of cancer of the thyroid gland was noted which can be connected with the improvement of the quality of diagnostics and the assignment of qualified specialists to the regions. The increase of the disease rate in other regions was not registered.

The haematological service is represented by a haematological department for 40 beds in an area hospital and 12 beds in a therapeutical department of a children hospital. In the regions there are no haematologists on the staff. In the area, there are 4 haematologist, they are all occupied.

The laboratory research is done by haematological, biochemical and cytological laboratories of area hospitals. Caryological research in genetic laboratories is not done.

According to the data obtained by the haematological department organized in 1987 the growth of haematological disease was noted, especially pronounced in 1988, in groups of acute leucosis, lymphosarcoma, erythremia, hypoplastic anaemia and other diseases.

In the western regions of the area, from 1987 to 1989, 13 cases of acute leucosis among adults were registered (5.3% of the total number of diseases) and 12 cases among children.

It is impossible to connect the growth of the number of cases of acute leucosis in the controlled regions with the radiation influence because caryological research was note done.

A considerable number of patients with thrombocytopenic purpura attracts attention, (in 1988 - 37 children, in 1989 - 39 children), but, from that number, only one child lives on the controlled territory.

Table 8. Structure of haematological diseases in the area (according to the data of the haematological department of the area hospital)

Year	Total No. of patients	Acute leucosis total w.regions	Chronic myelo-leucosis	Chronic lympho-leucosis	Myeloid disease	Erythremia
1987	409	44	33	76	27	18
1988	687	112	49	156	63	46
1989	560	90	40	82	51	38

Year	Lymthosarcoma	Myelo-fibrosis	Thrombocyto-penia	Anaemias total	hypoplastic
1987	2	20	22	90	14
1988	32	26	49	119	17
1989	20	20	31	100	9

The number of patients with hypoplastic anaemia remains stable from 1986 to 1989 (4 patients). The biggest number of iron-lacking anaemias was revealed among children in the Zlynkovsky region; in 1987 - 74, in 1989 - 120. In connection with this special attention should be given to the examination of children in other regions as well. The information about the structure of haematological diseases is given in table 8.

Table 9. Dynamics of the rate of infectious disease per 100,000 people from 1985 to 1989

Region	1985	1986	1987	1988	1989
Krasnogorsky	10,441	13,574	11,241	15,233	8,471
Gordeevsky	-	12,345	13,211	13,384	7,868
Novozybkovsky	55,256	46,278	39,850	58,820	48,894
Klintsovsky	12,336	12,271	9,273	16,095	11,233
Karachevsky (controlled)	17,139	27,419	11,429	12,153	11,773

The analysis of the rate of infectious diseases shows that in the controlled regions it has somewhat increased but the same increase was also noted in the controlled Karachevsky region, its highest rise occurring in 1988 (table 9).

The largest number of infectious disease was found in the Novozybkovsky region, which could be caused by the fact that Novozybkov is a town with a large number of workers in industrial enterprises who need sick time and leave to take care of a sick child. Other regions are village populations where the registration of such diseases is poorly conducted.

The analysis of the rate of infectious diseases dose not enable us to connect the radiation of the population with the lower immunity and thus to find the interdependence between the radiation and the change in the disease rate.

The study of the rate of the diseases of the cardiovascular system is a good illustration how the registered disease rate is connected with the quality of the conducted research and with the influence of the stress situation on the population. Thus, the rate of hypertensive disease found in the western and controlled regions shows that the clinical examination of the population, even with a simple measuring of arterial pressure, revealed the hypertensive disease in the controlled region two times more than in the period before continuous medical examinations and even by several tens of times more often than in the regions with a very low level of rendering medical aid in the pre-accident period (practically before the arrival of specialists from Moscow scientific research institutes). Thus, "the disease rate" in Brasovsky and Karachevsky regions has increased, in the Krasnogorsky region it has increased by three times, though did not reach the level of the controlled regions. It must also be noted that the number of patients is not large and it is not proper to make conclusions (table 10).

Table 10. Dynamics of the rate of hypertensive disease in 1988-1989

Region	Number of patients	
	1988	1989
Klintsovsky	15	35
Gordeevsky	35	61
Krasnogorsky	3	69
Novozybkovsky	79	185
Zlynkovsky	13	65

A similar explanation can be offered when analyzing the rate of cardiopathy and stenocardia. These indices may reveal hidden disease when electrocardiographic research is done, and if the doctor's attitude towards a patient is more considerate. One should also consider the small absolute quantities of patients revealed, concerning the calculations of indices, can depend on statistical fluctuations. However, one should not disregard the possible increase of the rate of cardiovascular disease caused by a constant stress situation.

HEALTH IMPROVEMENT

Certain measures are taken to improve people's health when disease are revealed during the clinical examinations (table 11).

When taking health-improving measures one needs to take into account the fact that there are no contraindications or specific requirements set for taking such measures and connected with living on a territory polluted with radionuclides. Patients should be sent to profile organizations for health-improvement, with vouchers being distributed and treatment recommended only by the doctor who conducted the clinical examination. It is permissible to use any medical and physiotherapeutical treatment, depending on the state of health. It is advisable to send people to health centers year round without any seasonal limits, taking the state of health into consideration and the fact that a rest is most efficient in the same climatic as the one in which a patient lives because sudden changes of climatic zones and adaptation when moving to the place of rest and returning home can have a negative influence, the real force of which has not been studies yet.

Table 11. Health-improvement among the examined population in the western regions in 1989

Region	Total No. people examined	Sick people revealed	Hospitalized	Out-patients' clinic treatment
Klintsovsky	3,710	1,998	251	1,998
Krasnogorsky	11,419	2,801	276	2,508
Novozybkovsky	51,063	15,628	14	8,822
Gordeevsky	11,793	4,149	1,069	2,586
Zlynkovsky	11,993	5,975	252	4,199
Total	89,973	30,551	1,862	20,113

CONCLUSION

The analysis of the disease rate in the controlled regions of the Bryansk area in 1989 did not allow for the establishment of the level of influence of radiation on the state of health of the population.

Further continuous observation of the state of health of the population and proper health-improvement of all sick people, especially children, are required.

In order to clinically examine the population living in the territories exposed to radiation fall-out, it is necessary to involve a great number of forces and resources, to computerize all CRHs and other organizations which participate in the medical examinations and to do a large amount of work training specialists to fulfil the clinical examination programs.

Presently, it is almost impossible to conduct a thorough medical examination of people living in the territory with a pollution density of up to 15 C per 1 km2 at the present level.

THE IMPACT OF LOW-DOSE IONIZING RADIATION ON THE PROGRESS AND OUTCOME OF PREGNANCY IN WOMEN

O.S. Ul'yanova, N.I. Mashneva
Institute of Radiation Hygiene, St. Petersburg

Analyzing the biological impact that low doses of ionizing radiation have on the reproductive ability of women and on the fetus is of considerable interest to both medical science and practical medicine. The studies devoted to this problems can be divided into two groups: the biological impact of radiation preceding conception and the effects of prenatal radiation. This work is devoted to the analysis of the latter problems.

Despite the importance of the problem, only a limited number of epidemiologic studies dealing with the implications of the prenatal radiation of humans exist [8]. For instance, M. Otake and M.J. Schull have assessed the mental development of children, (1251 in Hiroshima and 548 in Nagasaki), prenatally exposed to radiation at various gestation ages: 0-7 weeks, 8-15 weeks, 16-25 weeks, 26 weeks and more, and receiving doses ranging from 0.04 to 2.74 G. Researchers have shown that the highest probability of mental retardation was observed when radiation exposure occurred during the 8th to 15th week of gestation; the probability amounted to $4.5.ev^{-1}$, which is 4 times the case of radiation exposure after 16 weeks.

H. Ishikawa has described the pregnancy outcome of the women situated at a distance of 780-1180 m from the center of the explosion. 2 to 8 women advanced in pregnancy and experienced acute radiation sickness during pregnancy. Their newborn infants weighed from 1.3 to 2.8 kg, had microcephaly and suffered from deviations in intellectual and psychological development.

All of the aforementioned works deal with high-dose, acute irradiation. As far as the impact of the low dose is concerned, most works fall into the same pattern: they are built around an analysis of the influence of the x-ray inspections during the course of the pregnancy.

Based on the data collected in Finland's registry of congenital abnormalities, G. Cranroth has analyzed the connection between the specifics of the inspections during pregnancy and congenital abnormalities developed in children prenatally exposed to radiation. No risk enhancement has been detected in the case of stomach and pelvis radiation during pregnancy. In the case of chest radiation, the risk of CNS abnormalities increased by a factor of 2.2 in comparison to the control group.

K. Neumeister observed children under the age of 10 having received intrauterine radiation doses of up to 100 mGy during the course of x-ray inspections. cytogenetic analysis has not detected any variations. Of all the congenital abnormalities observed in the children, only hypospadia et spina bifida could be linked to radiation, since the time of radiation exposure coincided with organ differentiation.

According to the opinion of K.P. Il'yina, the characteristics of infants' physical development tends to change under the influence of fluorographical inspections of pregnant women [2].

Mary B. Meyer has shown that women prenatally exposed to radiation and receiving doses ranging from 1 to 5 rad tend to begin menstruation earlier, there is also an increase in the birth rate in women under 15 years of age, and an increase in the stillbirth and sterilization operation rates.

G.W. Kneale and A.M. Stewart have found that the risk of cancer incidence for children having received intrauterine x-radiation was higher than that for the unexposed ones; with the rates of risk dramatically rising with the age of the mother.

This work presents data concerning pregnancy and delivery outcomes together with the infants' state for a group of 306 women exposed to low doses of radiation as a result of the accident at the Chernobyl NPP and characterized by various gestation age at the moment of the accident. The main aim of this work was to clarify the relationship between the gestation age of the women and the corresponding fetal age (FTAG) at the moment of the accident and progress and outcome of the pregnancy, and physical development of the infants. For the analysis, all the pregnant women were divided into two groups according to the gestation age and the corresponding fetal age at the moment of the accident: those under 16 weeks, (that includes preimplantation, organogenesis and early fetal periods), 98 women, and those over 16 weeks (fetal period).

The gestation age has been calculated individually for each of the women using a formula proposed by the authors that involves the date of birth, length of pregnancy and the date of the accident: FTAG-PL - ((month of birth - 1) 30. + day of the month -116):7: where FTAG = the fetal or gestation age at the moment of the accident.

PL = pregnancy length counted in weeks.
30.5 = mean length of the month.
116 = number of days that passed from 1.01 to 26.04.86
7 = length of the week

The average overall external and internal dose, received by the inhabitants of the monitored area in 1986, has reached 9 mev. Since the progress of pregnancy is heavily influenced by the state of the thyroid gland function, the authors believe it is important to estimate the doses absorbed by the thyroid glands (TG) of pregnant women and fetuses.

Thyroid gland doses have been obtained using dosimetry data lists (form N 147 U). The authors have calculated the doses following instruction compiled by M.S. Balonov, T.V. Zhesko, A.I. Zvonova, Yu.O. Konstantinov and concerning the consumption of milk privately produced and brought in from the outside. The dose on the fetal thyroid gland has been calculated according to the formula proposed in the same instruction:

$$D_{fet} = K \times D_m$$

where K is the converting factor for the dose on the thyroid gland of the fetus; the factor depends on the gestation age. The average dose on the mother's thyroid gland amounted to 6.2±0.2 rad (from 2 to 22), the fetal gland, 10.2±1.6 (from 0 to 66 rad). In Figure 1 the authors present the dependence of the level of the dose absorbed by maternal and fetal glands, respectively, on the gestation age and FTAG at the moment of the accident.

Efforts aimed at calculating individual doses resulting from external exposure are still in progress.

The particulars of pregnancy and delivery outcomes have been studied by analyzing the individual case histories of the pregnant women, (accounting form 11), delivery histories, (ac.f. 9), the infants' development histories. The obtained data were processed using a DVK-2M computer, with the analysis involving statistical methods (Student t-test, x^2 criterion).

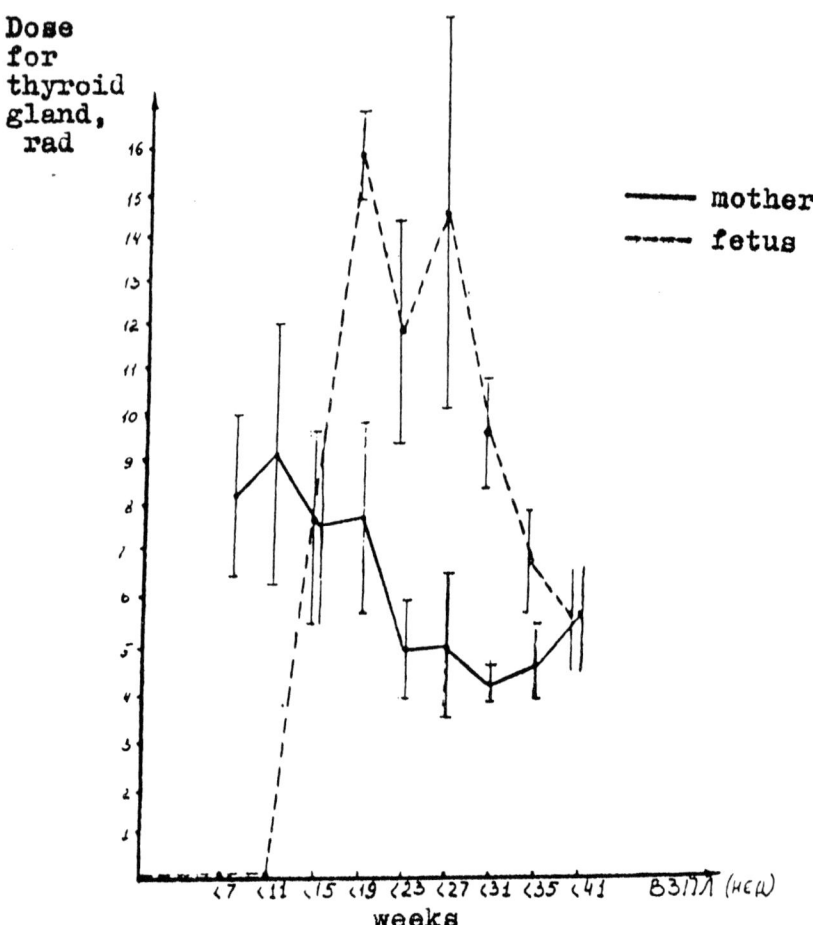

Figure 1. Dose Distribution for maternal and fetal thynoid gland

In order to facilitate drawing comparison between the groups under study, they were analyzed for a possible match in terms of age, fraction of primigravidi and primipari, start of menstruation and sexual activity, and a number of women with a complicated obstetrical history (COH), (spontaneous abortions, stillbirths, gynecological disease). From the data in table 1 it can be seen that no significant differences in the listed characteristics have been detected.

Table 1. General characteristics of the pregnant women

Characteristic	Gestation age		
	under 16 weeks N%	over 16 weeks N%	All the period N%
Age of the women	25.2±0.3	25.6±0.3	25.7±0.3
Age of their husbands	28.2±0.7	27.5±0.4	27.7±0.4
Start of the menses	13.8±0.1	13.7±0.1	13.7±0.1
Start of the sexual activity	20.4±0.3	20.4±0.2	20.4±0.2
Fraction of women with COH	26 26.5	47 23	73 24
primigravidi	42 43	82 39.4	124 40.5
primipari	45 46	90 43	135 44

62 characteristics describing pregnancy, delivery outcome, and the physical development of infants were analyzed within each of the groups; only 6 characteristics differ significantly for the two groups. 70.4% of women with a gestation less than 16 weeks at the moment of the accident had pregnancies complicated by some symptoms, (against 53.8%, P<0.1), and tended to have a higher rate of toxycosis during the second half-term. Among the symptoms of the late toxycosis the edema, or, as it was classified in the monitored area, hydrops gravidarum, predominated. Its frequency amounted to 36.7% for the women with early gestation ages at the moment of the accident, for those further advanced into pregnancy, to 20.7% (P<0.01). There are cases described in the literature, where a subclinical deficiency of TG function, detected only by functional analysis, resulted in a drastic increase in weight and edema.

The data of I.G. Randarenko [5] proves the existence of distinct moderate changes in the functional activity of TG in women whose pregnancy progress was affected by low-dose ionizing radiation, with the mean pregnancy dose reaching 10 mev, and a number of combined factors. For instance, a decrease in the thyroxin level has been detected. In addition, hydrocortizone is among the factors stimulating potassium, or water retention of the body. It is impossible to discount stress-related changes in the level of this hormone induced by a distrust to mass media and concern for their children.

In addition to this, women with gestation ages less than 16 weeks at the moment of the accident had a higher premature birth rate; i.e. their pregnancy fell short of 38 weeks, though their average pregnancy length did not vary.

No significant changes in the specifics of delivery, (early bursting of water, weakness of contractions, delivery-induced nephropathy and hypertension, manual removal of afterbirth caused by its adhesion of incomplete discharge), have been detected during the course of the year. Those with gestation ages under 16 weeks at the moment of the accident have been found to have a certain higher rate of trauma of soft maternal passages (27.6% of women against 16.8%, P<0.01).

Table. 2. Pregnancy Length

Characteristic	Gestation Age	
	Under 16 weeks N %	Over 16 weeks N %
mean pregnancy length	39.1±0.3	39.4±0.4
premature labor	22 22.5	24 11.5
pregnancy prolonged over 42 weeks	10 10.2	21 10.1

* = P<0.02

Table 3. Physical development and diseases of the neonates

Characteristic	Gestation age	
	under 16 weeks N%	over 16 weeks N%
mean weight	3383±60	3469±52
hypotrophy	15 15.3*	16 7.7
mass exceeding 4000 g	21 21.4**	33 15.9
mean length	51.2±0.4	51.5±0.2
mean Apgar score	8.3±1.0	8.6±0.9
asphixion	8 8.1	4 1.9
Breathing disorders	3 3.1	6 2.9

Footnote: * - P<0.1; ** - P<0.02; *** - P<0.01.

The data characterizing physical development of the infants did not reveal any significant difference in the mean values of birth weight and length. Still, an increase in the fraction of underweight infants, on the one hand, and those with weight exceeding 4000 g, on the other, has been detected only for women with gestation age less than 16 weeks at the moment of the accident. Infants in this group had certain lower Apgar scores (table 3). Still, detected deviations did not exceed the normal physiological limits.

Thus, as seen in the presented data, compared to those with a gestation age over 16 weeks at the moment of the accident, women with gestation ages under 16 weeks have a higher rate of complicated pregnancies and deliveries, and their infants were characterized by certain peculiarities in the physical development.

The authors believe, however, that the detected changes can not be unambiguously linked with the levels of ionizing radiation doses affecting pregnant women in the monitored area. Bear in mind that beside low doses of radiation, women are affected by a number of factors, not radiational in origin, (stress, nutritional disorders with vitamin deficiency predominant, hypodynamic, etc), that occur in the area under analysis. The listed factors affected the women with gestation ages less than 16 weeks for the longest period at the moment of the accident; their influence covered all the stages of pregnancy. It is generally agreed that these factors can considerably affect the reproduction function in their own right. For instance, emotional strain suffered during pregnancy often results in premature births [6] and pregnancy complications [4] and can lead to adaptationally hypotrophic changes in a fetal organism [3]. Hypodynamic induces a dramatic increase in the premature birth rate and can lead to an increase in the birth weight [1].

Outlined considerations lead us to conclude that changes detected in the reproductive ability or women affected by low-dose ionizing radiation are, apparently, induced by the combined action of various factors, radiational and non-radiational in origin. A final solution of the problem under consideration requires further research.

REFERENCES

1. **Bodashkin N.G.** et al. *Experimental clinical data concerning pregnancy and delivery progress under hypokinesia conditions* Akusherstvo i ginekologia. - 1988. N10. - p. 56-58.
2. **Il'yina K.P.** Concerning irradiation impact of x-ray inspections on the pregnant women *Radiational and hygienic assessment of radiational impact of x-ray inspections.* - Leningrad, 1972 - p. 72-73.
3. **Lyyutinsky S.I.** et al. Stress-induced hypotrophy and its correction by thymus preparations *Stress and immunity.* - Leningrad, 1989 - p. 30-31.
4. **Mirovitch D.Yu.** et al. On the gesthoses problem *Akusherstvo i ginekologia.* - 1988 - N12 - p. 58-60.
5. **Randarenko I.G.** Impact of the low-dose ionizing radiation on the pregnancy progress, delivery and neonate's state: *Brief of the candidate of medicine thesis.* - Leningrad, 1990 - 115 p.
6. **Sidel'nikova V.M., Sleptsova S.I.** Multivariate assessment of pregnancy non-preservation *Akusherstvo i ginekologia.* - 1988 - N6 - p. 18-20.
7. **Chetvertakov V.V.** et al. Implications of emotional stress for complications occurrence in obstetrical practice *Akusherstvo i ginekologia.* - 1988 - N6 - p. 17-19.
8. **USCEAR.** *United Nations Scientific Committee on the Effects of the Atomic Radiation: biological effects of prenatal irradiation.* - United Nations, New York, 1985.

Evaluation of Leukemia-Induction Risk Based on Analysis of the Consequences of Nuclear Incidents in the Southern Urals

M.M. Kosenko, M.O. Degteva, N.A. Petrushova
Institute of Biophysics, Moscow.

An investigation of the biological effects of protracted radiation is one of the principal trends of radiobiology and radiation medicine. Values of risk factor for somatic stochastic effects, (such as tumors and leukemia), following small dose protracted radiation have not been established yet [6,10].

A vast amount of information concerning the rate of leukemia evaluation in people exposed to radiation was obtained by a study of the contingents exposed to radiation in large doses, such as sacrifices of atomic bombardment in Japan, patients x-rayed for various medical indications and people affected while carrying out their professional duties [3,8,9].

Results of those studies provide the possibility to arrive at the conclusion that with doses above 0.5 Gy the risk of leukemia is most pronounced. As for smaller radiation doses, some authors emphasize a positive effect while others doubt it. Heated discussions are still in progress regarding the estimation of dosage rate efficiency value.

All that is required is collecting new data which might be obtained by prolonged observation of contingents exposed to protracted radiation.

This paper provides the results of long-term investigations concerning leukemia incidence and mortality rates among the inhabitants of the Southern Urals exposed to uranium fission products action.

In 1948, a defence industry enterprise engaged in plutonium production was put into operation in the Chelyabinsk region [4,5]. After the plutonium recovery at a nuclear chemistry plant, a significant amount of waste was left in the shape of a highly active radio-nuclides liquid which created a great deal of problems regarding their safe storage. Because of the fact that, at that time, there was no dependable technique for processing those wastes and for the disactivation of waste water, a flush discharge of those wastes containing an elevated percentage of radioactive elements into the Techa-Iset river system occurred. From 1949 to 1952, about 3,000,000 Ci of radioactive substances were dumped into the Techa river [7].

The unreliability of radioactive waste storage technique revealed itself in 1957 when, because of a tank heating and consequent chemical explosion, about 20 million Ci of radioactive substances were released into the environment [5].

Therefore, as a consequence of this plutonium producing activity, an unfavorable situation evolved in the Southern Urals in 50-s ensuing, from the radioactive substances dumped into the Techa river and, from the pollution of spacious territories after the chemical explosion in the radioactive waste storage tank (Figure 1).

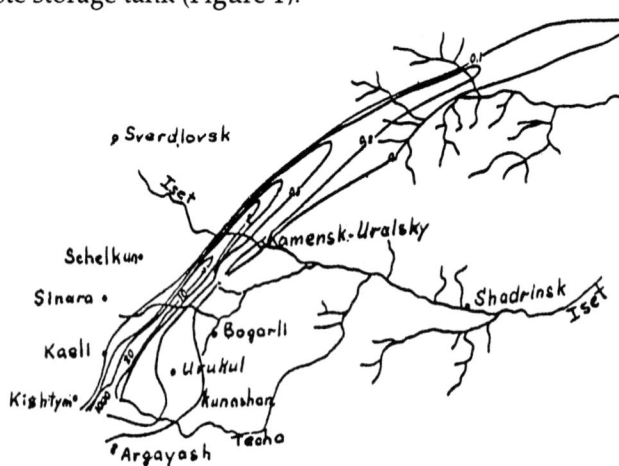

Figure 1 A sketch of radioactive polluted territories.

As for the isotope content of uranium fission products delivered into the Techa river, it comprised the radionuclide mixtures of strontium, cesium niobium, rutenium and rare earth elements. Approximately 25% of cumulative radioactivity was emitted from long-lived radio-nuclides: strontium 90 (T_{12} 29 years) and cesium-137 (T_{12} 30 years).

The Techa river is 240 km. long. A district administration center Brodokalmak was situated down-stream and 38 rural settlements. The total number of residents in the territory amounted to 28,000.

A radiation condition that developed there required the implementation of measures directed at lowering the radiation impact on the population. First, the dumping of radioactive substances into the river was greatly limited and afterwards cut down, then a prohibition on the use of river water for all household and drinking purposes followed, and then about 7.5 thousand people were moved out. As for the people yet there, to provide them with drinking and household water, some running-water mains as well as Artesian and pit-type wells were built. During contact with the Techa river the river-side population was involved in both external and internal radiation.

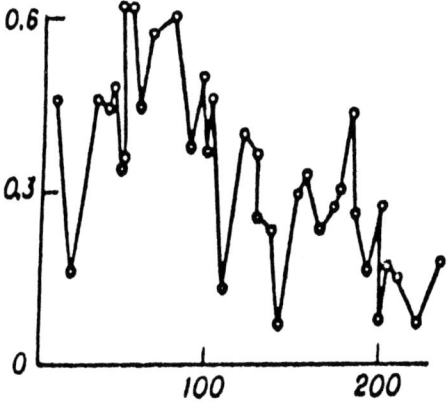

Figure 2. External radiation doses, taken against distance from radioactive fall-out

Axis of abscissas - distance (km) along the river
Axis of ordinates - dose absorbed, Gy

In 1951, the dosage rate of gamma radiation on the bank of Metlin pond, situated in the upper part of Techa river was in some places as high as 5 R h, on premises near the Metlino village, up to 3.5 R h, in streets and houses, up to 10-15 mR h. As early as 1952, a sharp decrease in the gamma dosage rate was observed, up to 50 mR h near the water line and 0.6 mR h in the territory of the settlement.

On the basis of measuring the gamma-radiation background value and assessment of people's behavior, the external radiation doses cumulated by the people of the Techa river were calculated. Figure 2 illustrates the change of the average doses cumulated from external radiation in relation to the settlement distance from the point of flush discharge.

The measures taken, (such as evictions, riverside enclosure, etc.), resulted in the cessation of the external radiation effect 6 years after its rise.

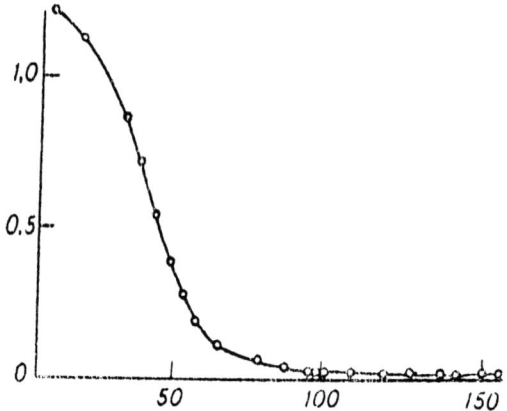

Figure 3. Average levels of internal red marrow dosage in Techa riverside residents.

Axis of abscissas - distance (km) along the river
Axis of ordinates - absorbed dose, Gy

In internal radiation, connected with the use of river water for drinking and cooking, the principal dose-forming radio-nuclide was strontium-90. In that event, the most affected tissues were the osteogenetic cell layer, covering the surface of the skeleton and red marrow. Measurements of strontium-90 content in the individual organisms were made by CU4-9.1 human radiation counter. To evaluate the internal radiation levels by the content of strontium-90 in a skeleton a specially developed model of that radio-nuclide metabolism was used [2]. Figure 3 demonstrates a change in the average internal radiation dose of red marrow in relation to the distance along the river. The cumulation of internal radiation doses continued for a prolonged period. The dose rate was greatest in the years following the gradual decrease. The limit of internal radiation dosage cumulation may be considered to be within the 20 to 25 year range, after that, the addition to the dosage was recorded as little as a few percent. Table 1 shows the average values of cumulative doses obtained from all external and internal radiation sources acting upon different organs and tissues of the Techa riverside residents. In the upper part of the river, the cumulative dose was found to be dependent upon external radiation but as one moves away from the point of fall-out radiation due to strontium-90, incorporated marrow tissue becomes predominant.

Table 1. Average doses absorbed by organs and tissues of Techa riverside residents and effective equivalent doses.

Settlement	Distance from fall out km	Doses absorbed by				Effective equivalent dose Sv
		marrow	bone surface	large intestine	other organs	
Metlino	7	1.64	2.26	21.40	1.27	1.40
Techa Brod	18	1.27	1.48	1.19	1.15	1.19
Asanovo	27	1.27	1.90	1.04	0.90	1.00
Nadyrovo	48	0.9	1.80	0.62	0.44	0.56
Muslyumovo	78	0.61	1.43	0.29	0.12	0.24
Brodokalmak	109	0.41	0.31	0.07	0.033	0.058
Russkaya Techa	138	0.22	0.53	0.10	0.037	0.082
Nizhnaya Petropavlovka	152	0.28	0.68	0.13	0.043	0.10
Shutikha	202	0.08	0.18	0.026	0.022	0.036
Zatecha	237	0.17	0.40	0.084	0.032	0.066

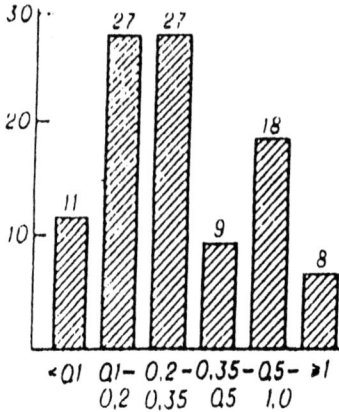

Figure 4. Distribution of radiation exposed population as per absorbed dose level in red marrow.

Axis of abscissas - absorbed dose range, Gy
Axis of ordinates - population per cent

Figure 4 demonstrates the distribution of the affected Techa riverside residents according to the dose levels found in red marrow. As seen in figure 4, about 40% of the people caught doses below 0.2 Gy, (small dose range), and 2,000 people (8%) had over 1 Gy dose.

Information about the radiation condition in the territory of the Eastern Urals radioactive track formed after the storage explosion in 1957 was reported at international meetings and published in the press [5].

In a fragments mixture following the 1957 explosion, a maximum percentage was recorded for isotopes of strontium-90, rutenium-106, cerium-144, and cesium-137. The territories contaminated with strontium-90 with pollution density over 200 Ci per Ikm2 was less than 1% of the track territory, while 94% of the territory was characterized by pollution density of 0.1 to 1.0 per Ikm2.

Radiation of the population comprised the following 4 components:

- external radiation from a passing cloud with maximum doses 1.3 mR;
- internal radiation induced by inhaling the moment the cloud passed, the dose rate on the lungs being within 0.05 Gy;
- external gamma radiation, induced by the fall-out of radionuclides;

- internal radiation, induced by penetration into the organism mixtures of fission fragments with food.

Effective equivalent radiation doses for the population were the following:

- for people evacuated in first 7 to 10 date (1054 persons)-0.52 Sv;
- for people evacuated in 250 to 330 days (6700 persons)-0.09 Sv;
- for those evacuated in 670 days (3100 persons)-0.03 Sv;
- for those not evacuated and left on the track territory;
- if pollution density was 1 to 4 Ci per Ikm2 for strontium-90-0.04 Sv;
- if pollution density was 0.1 to 1.0 Ci per Ikm2 for strontium-90-0.005 Sv.

The investigation of radiation-induced risk of leukemia among the exposed population of the Southern Urals has been carried out in Chelyabinsk branch No. 4 of the Russian Ministry of Health Institute of Biophysics.

With the object of people classification for the radiation induction level and for the correct evaluation of morbidity and mortality rate, a Register of people exposed was drawn up. The register is open for entry and is being replenished at the expense of the descendant generation born of people exposed to radiation. Information is logged in the register of people excluded from follow-up, in connection with death or migration to remote regions.

The Clinic department of the institute branch mentioned, undertakes the duties of a hematological center rendering medical assistance to adult hematological patients, residents of four rural administration districts of the Chelyabinsk region where radiation involved people dwell.

Information about mortality cases was obtained from the civil registrar's offices of the Chelyabinsk and Kurgan Regions. Death certificates were compared with the Register of radiated people and the lists were drawn up for both the groups affected and non-affected (control). The results were compared with those of two or three control groups which included people having no contact with radiation sources, although they lived in the same land areas.

The period of observation was 33 years, (from 1950 to 1982), for those exposed at the Techa river and 25 years, (1957 to 1982), for those living in the Eastern Urals radioactive track territory.

Data on leukemia incidence in people living in the Chelyabinsk region boundaries are presented in Table 2. For an investigation of the incidence dependence on the red marrow dose rate, the people exposed were subdivided into 5 dosage groups.

Table 2. Leukemia incidence rate for population irradiated on the Techa riverside.

Group	Average marrow absorbed dose Gy	No. of man-years in control period N	No. of leukemia cases n	Incidence rate n N per IO5 man-yr	90% confidence interval
observed					
1	1.43	46,777	6	12.83	5.58-25.31
2	0.82	36,339	3	8.26	2.25-21.33
3	0.59	90,909	9	9.90	5.17-17.28
4	0.29	106,808	12	11.24	6.48-18.20
5	0.13	108,047	7	6.48	3.04-12.17
control:					
1		1,956,000	133	4.50	3.88-5.20
2		1,664,000	93	5.59	4.63-6.64

Statistic correlation of the incidence rate with dosage rate evaluated by Pirson (r_p=0.77) and Spirmen (r_s=0.85) correlation factors proved to be valid at 95% level of confidence.

On the grounds of this data, that excessive leukemia morbidity is closely related to radiation effects, one can evaluate both the number of excessive leukemia cases as compared with the control groups and the time of their manifestation.

As seen in figure 5, a predictable leukemia morbidity rate rise in relation to both control groups was recorded in the 5 to 20 year range counting from the beginning of radiation, while most of the cases were recorded within the 15 to 19 year range. These data are in good accord with the now prevailing notions of radiation-induced leukemia manifestation time. An excess of leukemia cases in the follow-up period was but 23 cases with respect to control group 1 and 14 cases with respect to control group 2.

Leukemia-Induction Risk

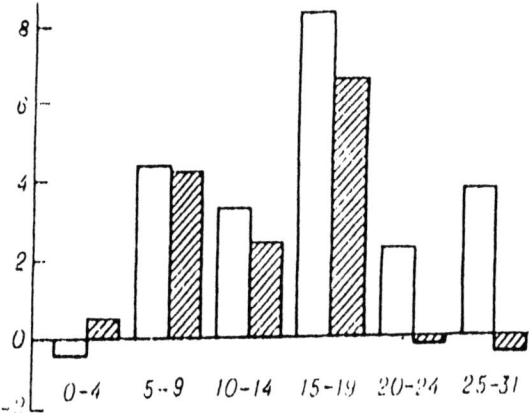

Figure 5. Dynamics of excessive leukemia incidents against control group No. 1 (white columns) and control group No. 2 (shaded columns).

Axis of abscissas - years after fall-out
Axis of ordinates - No. of leukemia excessive cases

Data on leukemia caused mortality cases in 6 dosage groups of Techa exposed people, and in 3 control groups, are indicated in Table 3.

Table 3. Leukemia mortality rate among Techa riverside residents

Group	Average marrow absorbed dose Gy	No. of man-years in control period N	No. of leukemia cases n	Mortality rate n N per 10^5 man-yr	90% confidence interval
observed:					
1	1.64	13,063	2	15.31	2.56-48.23
2	0.82	50,627	4	7.90	2.71-18.07
3	0.61	61,850	4	6.47	2.22-14.79
4	0.29	35,256	3	8.51	2.32-21.98
5	0.18	129,980	9	6.92	3.62-12.09
6	0.18	131,303	5	3.81	1.50-8.00
control:					
1	-	372,320	6	1.61	0.70-3.18
2		828,050	30	3.62	2.61-4.91
3		38.013	34	3.62	2.67-4.83

Some discrepancies in the quantity of the contingent studies for mortality and morbidity rates are explained by the fact that mortality analysis was made among the Techa riverside inhabitants, residing not only in Chelyabinsk but also in the Kurgan region.

Nevertheless, the incidence rate data proved to be more complete as some death certificates of leukemia-affected patients indicated as a main cause of death "Suicide" or "traffic accidents".

The most striking differences in the mortality rate among exposed and non-exposed people were observed from 1965 to 1974, i.e. after 15 to 25 years since beginning of radiation.

Table 4 gives data on the leukemia incidence rate among people who suffered in a 1957 nuclear accident.

Table 4. Leukemia incidence rate among population irradiated on the Eastern Urals radioactive track

Group	Average marrow absorbed dose Gy	No. of man-years in control period N	No. of leukemia cases n	Incidence rate n N per 10^5 man-yr	90% confidence interval
observed:					
1	0.20 -0.25	22,627	3	13.3	3.6-34.3
2	0.03 -0.12	112,900	9	8.0	4.2-13.9
3	0.003-0.03	357,940	23	6.4	4.4- 9.1
control:	-	526,700	28	5.3	3.8- 7.3

The leukemia morbidity rate was found to be the highest in the first group but so far, only 3 cases were recorded, a discrepancy with control group was statistically negligible.

An assessment of the radiation-induced risk of leukemia was accomplished on the basis of the following:

- model of absolute risk which is considered judging by references most preferable for leukemia cases [11,12];
- linear dose-effect correlation recommended by the UN organization for intermittent ionizing radiation limited by upper levels of the order of 1 Gy.

Figure 6 demonstrates regression lines for morbidity and mortality rates.

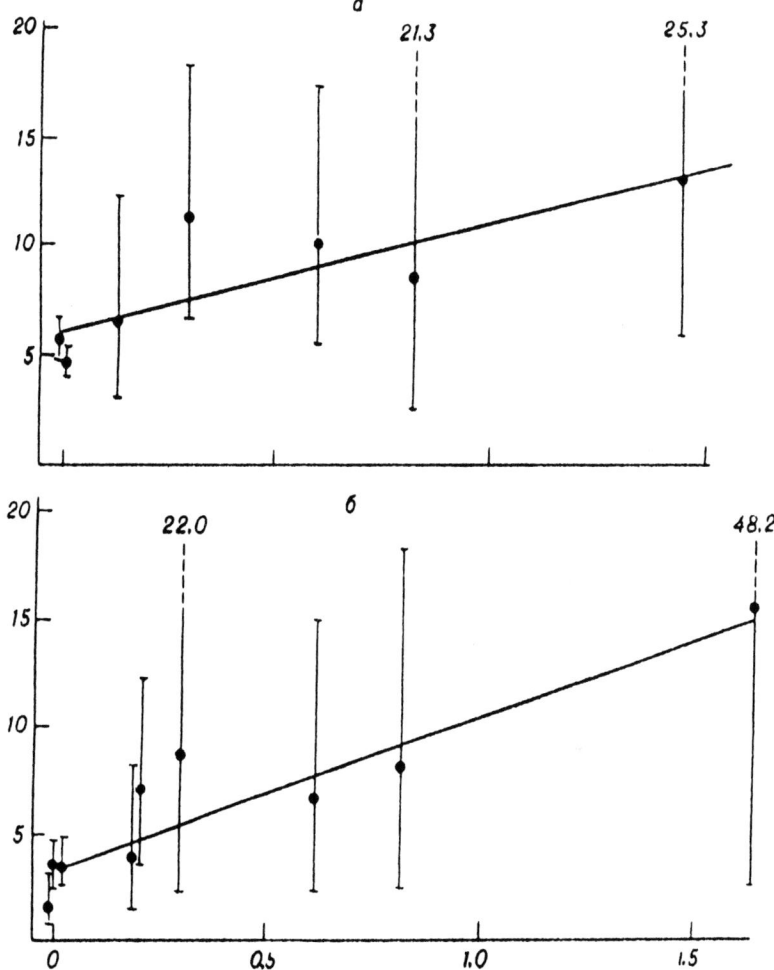

Figure 6. Leukemia incidence against red marrow dosage estimated after incidence (a) and mortality (b) data

Axis of abscissas - absorbed dose, Gy
Axis of ordinates - leukemia incidence, 10^5 yr 1

In Table 5, a variety of evaluation methods for leukemia risk assessment (R) for Techa population are summarized. Apart from calculations in figure 6, curves, Table 5 comprises the results of calculations made by excessive incidence dynamics (see figure 5) as well as by linear approximation of excessive incidence in dosage points. In the last event the line slope angle was estimated in consideration of "weighting" each dosage point, which was inversely proportional to the width of confidence interval.

Table 5. Estimation versions of leukemia risk factor based on Techa population data

Estimation method	Data used	Risk factor R per 10^4 man-yrs
By excessive No. of cases: R=n N.D, where N-No of man-years, D-average marrow absorbed dose	N=388 900 man-yrs D=0.54 Gy Excessive No. of cases control group 1 - n=23 control group 2 - n=14	1.10 0.67
By slope angle of regression line for leukemia cases (Y) against dosage (D) $Y(D)=Y_o+R.D$ where Y_o, R - parameters estimated.	Incidence rate in 5 dosed and 2 control groups rating of $Y=6.0 \cdot 10^{-5}$ Mortality rate in 6 dosed and 3 control groups (rating $Y_o=3.6 \cdot 10^{-5}$)	0.48 0.68
By slope angle of weighted regression line for excessive leukemia incidences (Y) against dosage (D) with fixed zero point: Y(D)=R.D. where R-parameter estimated, weight factors are inversely proportional to width of confidence intervals.	Excessive incidence rate in 5 dosage points against control group Excessive mortality rate in 6 dosage points against control groups compatible in national and residence aspects	0.68 0.79

As seen in Table 5, all variants of calculation providing similar results and range of risk factor estimations was within R=0.48-1.10 per 10^4 person-Gy.

As seen in Table 6, both the scope and character of the study provides the possibility of comparing the results with the data put in the basis of risk factor estimations included in the documents of the UN Scientific Committee on Effects of Atomic Radiation.

Table 6. Main character of data taken for estimation of leukemia risk factor

Parameter	Persons survived after bombardment [13]	Patients affected by anchylosis spondylite [10]	Females irradiated for uterus cervix cancer (J. Boice data)	Techa riverside residents
No. of affected thousand	42	14	83	27.8
Females percent	59	17	100	56
Age in time of exposure	0-90	15	30-70	0-90
Control type	Internal 34,000 people	average over country	internal and average over country	internal and average over region
Dose meter type	Individual DS86	Individual and average group	Average for settlement	
Character of irradiation	Single external, even	External fractionized uneven	Chronic external uneven	Chronic external and internal from radionuclides Sr and Cs
Average dose on red marrow, Gy	0.24	1.9	7.5	0.40
Range of individual doses Gy	0.01-6.0	0-8.06	3.0-13.0	0-3.0
No. of man-yrs under risk, thousands	1134	174	624	422
Absolute risk factor per 10^4 man-Gy	2.94 (2.43-3.49)	2.02	0.61	0.48-1.10

The authors rating of the risk factor appears to be a little lower than the results of Y. Shimizu et al. [13] for Hiroshima and Nagasaki dosage re-estimation or of S. Darby et al. [10] for patients suffering from anchylosis spondylitis disease, but are compatible with the data of J. Boice on female patients radiated for uterus cervix cancer.

It is worth mentioning that the last case presents some complexities regarding the determination of radiation effect since dissipation of the absorbed doses by the organism was extremely uneven. Risk factors (levels) obtained by Y. Shimizu and the authors differ as much as 3 to 7 times. As for the Techa population dose rate, the affecting red marrow did not exceed 0.05 Gy a day, these differences may be regarded as an effect of the dosage rate.

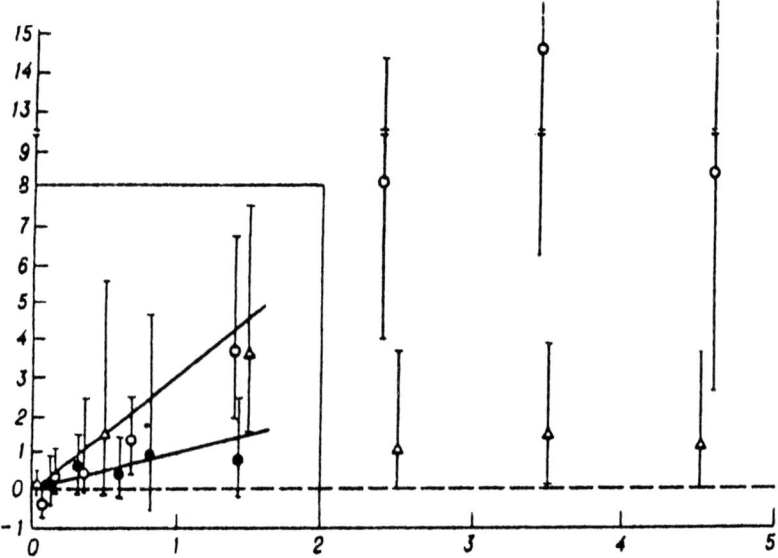

Figure 7. Excessive leukemia incidence against red marrow dosage after different author's findings

Legends: black circles - author's own data
white circles - Y. Shimizu data
white triangles - S. Darby data
Outlined sector bounds area of 1 to 3 cases per IO4 man-yr
Axis of abscissas - absorbed dose, Gy
Axis of ordinates - leukemia excessive incidence IO-4 yr

It might be of interest to correlate the excessive leukemia incidence observed in some dosage points with results of other studies (figure 7). As seen here, in a dosage range below 1.5 Gy most of the points are found in the area bounded by a sector of 1 to 3 excessive cases per 10^4 people per 1 Gy a year. The presented estimations give an idea of the actual consequences of people exposed to small doses of ionizing protracted irradiation.

Acknowledgement

The authors are grateful to L.A. Ilyin, A.K. Guskova, I.V. Filyushkin, M.M. Saurov, G.M. Avetisov, L.A. Buldakov and O.A. Pavlovsky for participation in discussions concerning the results of this study.

References

1. **Belle, Yu.S., A.N. Kovtun** et al. *Med. Radiol.*, 1975, Vol 20, No. 6, pp. 52-58
2. **Dyogteva,M.O. and V.P. Kozheurov** "Influence of age on radiation effect levels during radiostrontium penetration into the human organism". *NKDAR A AC Document.* 82 G L. 1768, Moscow, 1989
3. "Sources and effects of ionizing radiation". *NKDAR*, UN. Report, 1977, Vol. 2, New York, 1978
4. **Nikipelov, B.V., G.N. Romanov,L.A. Buldakov** et al. *Atom. Energiya*, 1989, Vol. 67, No. 2, pp. 74-80
5. **Nikipelov, B.V., A.F. Lyzlov and N.A. Koshurnikova** *Nature*, 1990, No. 2, pp. 30-38
6. "Antiradiation protection". *ICRP Recommendations*, Publication No. 26, Moscow, 1978
7. **Ryabyev, L.D.** *Chelyabinsk. Rabochiy*, 1989, 6 July, p. 1
8. **Beebe, G.W.** *Amer. J. Epidemiol.*, 1987, Vol. 144, No. 6, pp. 761-783
9. **Darby, S.C., S. Nakashima and H. Kato** "Radiat. Effects Res. Found".: *Techn. Rep.*, 1986, Vol. 4, No. 84, pp. 1-128
10. **Darby, S.C.** *Hlth Phys.*, 1986, Vol. 51, No. 3, pp. 269-281
11. **Gilbert, E.** "Health Effects Model for Nuclear Power Plant Accident Consequence Analysis", New York, 1985
12. **Pochin, E.E.** *Brit. J. Radiol.*, Vol. 60, pp. 42-50
13. **Shimizu, Y., H. Kato, W.J. Schull** et al. *Radiat. Res.*, 1989, Vol. 118, pp. 502

SUBJECT INDEX

—A—

acute radiation syndrome, 30, 48, 51, 76, 77
alpha emitter, 32
Arterial Blood Pressure, 18

—B—

Belorussia, 34, 35
bone marrow, 31, 37

—C—

cancer, 19, 21, 24, 25, 27, 31, 32, 49, 69, 70, 71, 73, 103, 110, 129
cancer of the esophagus, 25
cardiovascular system, 19, 105
Chelyabinsk, 15, 22, 26, 118, 123, 124, 126, 131
Chernigov, 78, 80, 85, 86, 88
Chernobyl, 1, 2, 3, 5, 10, 13, 29, 53, 61, 75, 76, 79, 81, 83, 86, 90, 91, 110
Cheyabinsk province, 26
chromosome anomalies, 22
chronic exposure, 30, 78
chronic radiation exposure, 37, 78
Chronic Radiation Syndrome, 47
circulatory system, 22
congenital anomalies, 10, 21
Congenital Developmental Anomalies, 21
contaminated land, 17
contamination density, 7, 15, 17, 18, 34
contamination levels in agricultural produce, 17
cytogenetic effect, 76, 77, 80
cytogenetic markers, 77, 78

—D—

Decontamination, 17
Dosage monitoring, 40
dose levels, 31, 33, 34, 46, 71, 73, 122
dosimeters, 40
dynamics of children's morbidity, 7

—E—

Evacuation, 17, 18, 84
Exposure Dose, 18
external gamma exposure, 33, 48, 49, 50

—F—

Fertility, 56, 58

—G—

gamma emitters, 15
gamma radiation, 39, 40, 47, 49, 50, 120, 122
gamma radiation exposure, 40, 47
genetic mutations, 75
Gomel', 35

—H—

heart disease, 24

—I—

infant mortality, 21, 23, 34, 35
Infants, 22, 23, 115
iodine, 78, 83, 84, 85, 86, 95, 101
iodine data bank, 84, 85
ionizing radiation, 5, 21, 39, 48, 53, 58, 75, 78, 109, 113, 115, 116, 126, 131
iron deficiency anemia, 36
iron deficiency anemias, 36
irradiation accident, 26
Ivankovsky, 7

—K—

Kiev, 5, 7, 8, 9, 10, 13, 75, 83, 85, 86, 88
Kiev region, 5, 7, 9, 11, 13

Kozelets, 78, 79, 80
Krasnopol'e district, 35, 36

—L—

leukemia, 25, 37, 69, 117, 118, 123, 124, 125, 126, 127, 128, 129, 130
leukocyte counts, 19
leukopenia, 20
local radiation injuries, 30
Luginsky, 5, 6, 7, 8, 10
Luginsky district, 10
lymphatic neoplasms, 25
lymphopenia, 32

—M—

Maligant Neoplasms, 25
malignant growths, 47, 48
malignant neoplasms, 10, 26, 99
malignant tumor mortality levels, 49
malignant tumors, 24, 31, 50, 89, 90, 102
Mayak, 30, 31, 33
medical demographic indices, 35
medical surveillance, 19, 40
Mogilev, 35, 36
mutagens, 78, 81

—N—

Narodichi, 5, 6, 7, 9, 10, 79, 80, 81
neural system, 37
neutrophil formula, 20
Norodichi, 9, 78
Nuclear facilities, 50
nuclear reactor, 39, 41, 83
nuclear weapons, 29, 39, 40, 50, 75

—O—

occupational exposure, 33, 41, 45, 46, 47
Olevsky region, 86
oncologic morbidity, 61, 67, 68, 69, 70, 71, 72, 73
oncologic mortality, 67, 68, 69, 70, 72
oncological mortality, 32, 49, 50
oncological pathology, 31
osteochondrosis, 20
otolaryngeal examinations, 19
Overexposure, 47
Ovruch, 5, 6, 7, 8, 9, 10, 78, 80, 81, 86

—P—

pesticides, 13, 80
Polessky, 7
Polessky district, 7, 8, 9, 10
population mortality rate, 32
pregnancy, 34, 57, 58, 59, 109, 110, 111, 113, 114, 115, 116
Pripyat', 85, 86

—R—

radiation sickness, 2, 19, 47, 48, 109
radiation waste, 33
radioactive contamination, 5, 8, 9, 10, 13, 27, 41, 45, 71
radioactive decay, 15
radioactive fall-outs, 76, 77
radioactive wastes, 31
radioactivity, 84, 119
radiochemical facility, 51
radiochemical plant, 41, 46, 48, 49, 83
radioiodine content, 36
Reorganization of agriculture, 17
reproductive functioning, 53

—S—

Semipalatinsk test site, 37
Serum cholesterol levels, 19
Sham irradiation, 54
somatic problems, 19
stress, 36, 48, 53, 54, 55, 56, 57, 58, 73, 95, 105, 106, 113, 115, 116
strontium-90, 16, 120, 121, 122, 123
Sverdlovsk, 15, 22, 26

—T—

thyroid doses, 36
thyroid gland, 2, 31, 35, 83, 84, 95, 101, 102, 103, 111
thyroid hyperplasm, 36
Thyroidal Radiation Doses, 83
trithium, 33
tritium, 33
tumors, 24, 26, 31, 37, 50, 89, 90, 102, 117
Tyumensk, 15

—U—

Ukraine, 2, 5, 6, 7, 8, 9, 10, 13, 75, 77, 78, 80, 81, 83, 85, 86
underground nuclear tests, 37
uranium fission products, 37, 118, 119

—Z—

Zhitomir region, 5, 7, 78